Electric Vehicle Industry

Insights and Perspectives

Series

Tech Industries — Insights and Perspectives Series

Mustafa Al-Dori

Copyright Page

Copyright © 2024 Mustafa Al-Dori. All rights reserved.

Dedication

For my parents, whose care, patience, and guidance helped shape who I am.

Acknowledgments

My sincere thanks to everyone who supported this book—my family, friends, mentors, and colleagues—for their encouragement and guidance. Thank you for being part of this journey.

Disclaimer

This book is for education only. It explains electric vehicles as systems and as an industry. It does not give legal advice. It does not replace official rules, standards, or professional guidance. It does not provide engineering design instructions for building or modifying vehicles. It also does not provide investment advice.

Electric vehicle laws, safety rules, and incentives can change by country and time. Your situation may be different from the examples in this book. If you need legal or compliance guidance, speak with a qualified professional. If you need engineering decisions, use official standards and certified experts. If you need financial decisions, speak with a licensed advisor.

You are responsible for how you use information in this book. The author and publisher are not responsible for losses, damages, or outcomes that may result from using this content.

About the Series

Tech Industries — Insights and Perspectives Series is built for readers who want clear understanding without hype. Each book explains one industry as a system. The goal is practical literacy. The goal is not noise, trends, or marketing language.

The series uses the same method in every title. It starts with a simple mental model. It then adds the technology layer, the market layer, and the risk layer. It ends with future trends using scenarios, not fantasies. You learn how to think, not what to repeat.

The writing style is global and human. It is designed for readers at B1–B2 level. Technical terms appear only when needed. Each term is defined the first time it is used.

About This Book

Electric vehicles are not just "cars with batteries." They are an energy system on wheels. They connect batteries, power electronics, software, charging, and the grid. They also connect supply chains, regulation, and business models.

This book gives you a structured map of that world. It explains what matters, why it matters, and how parts connect. It shows trade-offs in simple language. It also shows where headlines can mislead you.

You will learn the core building blocks of EVs. You will learn what changes range and battery life. You will learn why charging feels easy in one place and hard in another. You will also learn how companies compete, and where value is captured.

How to Use This Book

Read this book like a system guide. Start by learning the main components and their relationships. Then move to the technology that makes the system work. After that, study charging and the grid, because the car does not live alone.

If you are new, do not rush into advanced chapters. Build your base first. If you already work in the field, use the check sections to test your thinking. If you are a business reader, focus on value chains and adoption drivers.

Keep a simple note while you read. Write down terms you want to revisit. Write down trade-offs that surprised you. Then return to the glossary when a word feels unclear. This method keeps your learning calm and steady.

Reading Paths (3 Routes)

Route 1: Fast Clarity Path

Start with the Introduction and Part 1. Then read the chapters on batteries and charging. End with markets and future trends. This path builds a clear mental model quickly.

Route 2: System Builder Path

Read from Part 1 through Part 4 in order. Then read the grid and charging network chapters carefully. End with lifecycle and future scenarios. This path is best for deep understanding.

Route 3: Business and Strategy Path

Read the Introduction and Part 1. Then move to value chains, pricing, and adoption drivers. Then read regulation and safety at a high level. End with future trends and scenario thinking. This path is best for decisions and planning.

Quick Wins Map (1 Page)

Use this page when you want clarity fast. Read it before meetings or before buying decisions.

First, define your goal in one sentence. Then ask what "good" means for that goal. Range is not always the main target. Charging access can matter more than battery size.

Next, separate marketing claims from measurable factors. Ask about real driving conditions. Ask about charging curve behavior, not peak numbers. Ask about warranty terms and service support.

Then check the system outside the car. Home charging changes everything. Public charging reliability also changes everything. Grid prices and time-of-use plans can change cost outcomes.

Finally, apply one safety habit. Treat batteries as energy-dense devices. Avoid heat abuse and poor charging habits. Follow manufacturer guidance for storage and charging.

ELECTRIC VEHICLE INDUSTRY
Insights and Perspectives

Educational only

EVs are not just cars with batteries. They are an energy system on wheels.

Energy	Powertrain	Control	Charging	Grid	Lifecycle
Stored energy and condition over time.	Turns energy into motion	Software manages power and safety	How energy returns (home or road)	External constraints, capacity, and cost	

If one link is weak, the whole experience suffers.

Common mistake
People debate one link only (range, charging speed, or battery cost)

QUICK WINS

 Goal
Define your goal in one sentence. Ask what 'good' means.

🔍 **Measurable vs marketing**
Separate claims from measurable factors. Ask about real conditions.

 The world outside the car.
Home charging and public reliability change everything.

 One safety habit
Treat batteries as energy-dense devices. Avoid heat abuse.

— 3 READING ROUTES —

Intro + Part 1 → Batteries + Charging
———→ Markets + Future. ———

WHERE VALUE IS CAPTURED

Value is create and captured in many places, not one.

— 3 READING ROUTES —

⚡ **Fast Clarity**
Intro + Part 1 → Batteries + Charging

 System Builder
Part 1-4 in order → Grid + Charging network → Lifecycle + scenarios

— WHERE VALUE IS CAPTURED —

Value is created and captured in many places, not one.

Educational only. Lows and incentives change by country and time. Use this as a thinking tool, not a promise.

Table of Contents

Introduction	12
Part 1 — Foundations Without Confusion	16
Chapter 1 — What an Electric Vehicle Really Is (In Plain Engineering English)	17
Chapter 2 — A Short History of EVs (What Changed, and Why)	24
Chapter 3 — EV Types and Use Categories (Not Just Body Styles)	29
Part 2 — The Technology Layer	33
Chapter 4 — Battery Basics for Non-Specialists	34
Chapter 5 — Motors, Inverters, and Power Electronics	39
Chapter 6 — Thermal Management and Safety	43
Chapter 7 — Charging Hardware and Charging Speed Reality	47
Chapter 8 — Software, Controls, and the EV "Digital Layer"	50
Part 3 — Charging Networks and the Grid	53
Chapter 9 — Charging Networks as Infrastructure	54
Chapter 10 — Grid Impact and Smart Charging (High Level)	60
Chapter 11 — Vehicle-to-Home and Vehicle-to-Grid (V2H/V2G)	65
Part 4 — Design, Manufacturing, and Quality Systems	71
Chapter 12 — EV Platform Design and Packaging	72
Chapter 13 — Battery Manufacturing and Supply Constraints	78
Chapter 14 — Assembly, Testing, and Reliability	84
Part 5 — Markets, Business Models, and Competition	91

Chapter 15 — The EV Value Chain (Who Makes Money Where)	92
Chapter 16 — Pricing, Total Cost of Ownership, and Adoption Drivers	98
Chapter 17 — Competition, Differentiation, and Brand Trust	104
Part 6 — Regulation, Safety, and Responsible Adoption	110
Chapter 18 — Regulation and Standards (Global View, High Level)	111
Chapter 19 — Safety, Risk, and Public Trust	118
Part 7 — Sustainability, Materials, and Lifecycle	124
Chapter 20 — Materials, Mining, and Supply Ethics (High Level)	125
Chapter 21 — Lifecycle Thinking and Carbon Accounting (High Level)	130
Part 8 — Future Trends (Practical Forecasting)	135
Chapter 22 — Future of Electric Vehicles (2026–2035)	136
Conclusion	142
Glossary	144

Introduction

Electric vehicles are moving from the margins to the mainstream. This shift is not only about new cars. It is about a new way to store and use energy in daily transport. It connects engineering, software, infrastructure, and policy in one system. That is why EVs can feel simple on the surface, but complex underneath.

This book is written to remove confusion without adding noise. You will not find hype here. You will also not find fear-based claims. You will find calm explanations, clear trade-offs, and practical mental models. When something depends on place or time, we say so. When something is uncertain, we label it as uncertain.

Why Electric Vehicles Matter Now (Without Noise)

EVs matter because transport is a large part of modern life. People need mobility for work, family, and business. EVs offer a different technology path for that mobility. They can reduce local tailpipe emissions during driving. They can also change energy costs and maintenance patterns. The exact benefits depend on the electricity mix, the vehicle design, and the charging setup.

EVs also matter because they force new infrastructure decisions. Charging is not only a "feature." Charging is a system that must work reliably. Grid capacity, charger availability, pricing, and standards can shape the real experience. In many cases, the charging experience becomes as important as the vehicle itself.

EVs matter to businesses for another reason. They reshape supply chains and competition. Batteries, software, and power electronics become central. New companies appear. Old companies adapt. Some changes are visible to consumers. Many changes happen behind the scenes in manufacturing and sourcing.

What This Book Is (and Isn't)

This book is a structured guide to understanding the electric vehicle industry. It explains EVs as systems, and it explains the industry as a value chain. It focuses on concepts that stay useful even when specific products change. It aims to give you vocabulary, mental models, and clear thinking tools.

This book is not a buying guide for a specific model. It is not a repair manual. It is not a legal document. It is not an engineering design standard. It is also not investment advice. You should use official standards and qualified professionals for decisions that require formal responsibility.

A Simple Mental Model: Energy → Powertrain → Control → Charging → Grid → Lifecycle

Think of an EV as a chain of connected parts. Energy is stored, converted, controlled, and replenished. Each link affects the next link. If one link is weak, the whole experience suffers.

Energy is the battery and its condition over time. Powertrain is the motor, inverter, and drivetrain that turn energy into motion. Control is the software layer that manages power, safety, and driver demands. Charging is how energy returns to the battery, at home or on the road. Grid is the external system that supplies electricity and sets constraints and prices. Lifecycle is the full story from materials and manufacturing to use, reuse, and recycling.

This model helps you avoid one common mistake. Many discussions focus on only one link. People focus only on range, or only on charging speed, or only on battery cost. Real understanding comes from seeing how the links trade off against each other.

How to Read This Book Like a System

Start by building a stable base. Learn what the main parts are, and what each part does. Then learn the trade-offs that connect those parts. After that, move to the outside system. Charging networks and the grid shape daily reality more than many headlines suggest.

Read actively. When you meet a new term, pause and define it once. When you see a claim, ask what it depends on. Ask what can change the outcome. This habit keeps you calm and precise. It also protects you from oversimplified stories.

If you are new, follow the order of Parts. If you are experienced, use the chapters as a reference map. You can jump to the section you need, then return to the system view. The goal is not memorization. The goal is reliable thinking.

Quick Check: 7 Questions to Keep Your Thinking Clear

1. What problem are we solving: cost, emissions, performance, convenience, or compliance?
2. Which link is the constraint: battery, charging, grid access, software, or service support?
3. Are we discussing energy (kWh) or power (kW), and are we mixing them by mistake?
4. Are the claims based on ideal conditions, or on real driving and real weather?
5. What depends on location: electricity price, charging availability, rules, and grid capacity?
6. What is the main risk: safety, reliability, cost surprises, or operational downtime?
7. What would change your conclusion: better charging access, different use case, or different vehicle class?

Part 1 — Foundations Without Confusion

Electric vehicles can look simple from the outside. You plug in, you drive, and you charge again. The confusion starts when people explain EVs with one metric only. Some people reduce everything to range. Others reduce everything to charging speed. Others reduce everything to battery chemistry. The truth is more connected than that.

In this part, we build a clean mental model. We treat the EV as a system with clear links. Each link has limits and trade-offs. Each link also creates business consequences. When you understand the links, you stop chasing headlines. You also stop comparing vehicles with the wrong questions.

You will learn what an EV really is, in plain engineering English. You will see the main subsystems and how they interact. You will also learn why EV history matters today. History explains why the current wave looks different. It also explains what still does not change.

Finally, you will learn how to classify EVs by job, not by shape. A sedan, a bus, and a delivery van can share EV logic. Yet they face very different constraints. The right choice depends on mission, environment, and infrastructure. This part gives you that decision clarity.

Chapter 1 — What an Electric Vehicle Really Is (In Plain Engineering English)

An electric vehicle is a vehicle that uses an electric motor for motion. It stores energy in a battery and converts that energy into movement. That sounds simple. The details matter because the battery is not a fuel tank. A fuel tank stores energy passively and releases it through combustion. A battery stores energy chemically and releases it through controlled electricity. That difference changes design, cost, safety, and daily use.

A conventional car has three major ideas at its center. It has a fuel system, an engine, and a transmission. The engine turns fuel into mechanical power using heat. The transmission keeps the engine in a useful operating range. The system works well because fuel has high energy density. Refueling is fast and widely available. The trade-off is complexity, heat, noise, and more maintenance. The trade-off also includes tailpipe emissions during driving.

An EV replaces the engine with an electric drivetrain. The core conversion happens in the motor and the inverter. The inverter controls how electricity flows to the motor. The motor then produces torque with high precision and fast response. This is why EVs often feel smooth. It is also why traction control can feel more direct. The vehicle can adjust torque quickly and repeatedly.

The battery pack is the center of the EV from a system view. It is the largest energy store and a major cost driver. It is also a major weight driver. The pack contains many smaller units called cells. Cells are grouped into modules in many designs. The pack includes sensors, wiring, cooling channels, and safety components. A battery pack is not only "more battery." It is a managed energy device that must stay within safe limits.

The battery management system, often called the BMS, is the pack's brain. The BMS monitors voltage, current, and temperature. It also estimates the battery's state of charge. State of charge is the system's best estimate of how full the battery is. It is an estimate because batteries do not behave perfectly. Temperature and aging can change behavior. The BMS also tracks state of health. State of health describes how the pack's usable capacity compares to new. It declines over time due to degradation.

Degradation is the gradual loss of battery performance and capacity. It is not one simple process. It is the result of chemical changes and stress over many cycles. Heat is a key driver of stress. Very high or very low temperature can hurt performance. Fast charging can also add stress in some conditions. Deep cycling can add stress in some conditions. The point is not to fear degradation. The point is to understand that batteries age like other components. They need reasonable care and realistic expectations.

The power electronics layer is another core subsystem. Power electronics includes the inverter and other converters. A converter changes voltage levels for different needs. The motor needs controlled power delivery. The 12-volt system still exists in most vehicles. Lights, computers, and accessories rely on it. A DC-to-DC converter often manages that supply. This is one reason EVs have strong software needs. The system must coordinate energy flow across several voltage levels.

Thermal management is a subsystem many beginners underestimate. Batteries and power electronics operate best within a temperature window. Motors also create heat under load. The vehicle therefore needs cooling and, in some climates, heating. Thermal management is not comfort air conditioning. It is a performance and safety layer. It influences charging speed, sustained power, and battery aging. In cold conditions, the pack may need warming for fast charging. In hot conditions, the pack may need stronger cooling under load.

Charging is the EV's refueling layer. Charging is both an on-board and off-board system. The grid provides electricity in alternating current for most homes. The vehicle must convert it to direct current for the battery. The on-board charger handles that conversion for AC charging. For fast charging, many stations provide direct current. The station then controls the power flow and the vehicle coordinates acceptance. This is why charging speed is not only the charger's rating. It is also the vehicle's battery design, temperature, and control strategy.

Range is the EV metric most people ask about first. Range is the distance a vehicle can drive on a charge. Range depends on the battery's usable energy and the vehicle's efficiency. Usable energy is often lower than the pack's total size. The system keeps buffers for safety and longevity. Efficiency depends on speed, temperature, wind, and driving style. Highway driving often uses more energy per distance. High speed increases aerodynamic drag strongly. Cold weather can also reduce effective range because heating needs energy and batteries behave differently when cold.

Power is the vehicle's ability to accelerate and sustain speed. Power depends on motor capability, inverter limits, and battery discharge limits. A vehicle can have high peak power but limited sustained power. Heat can limit sustained power in some conditions. This is why performance claims must be read carefully. A short burst is different from repeated hard use. The same is true for fast charging. A peak charging number is not the full story. The charging curve usually tapers as the battery fills.

Cost is shaped by the battery and the supply chain behind it. EVs also contain high-value electronics and software. Some EVs reduce mechanical complexity compared to combustion cars. They often have fewer moving parts in the drivetrain. That can reduce some maintenance types over time. It does not remove all maintenance. Tires can wear faster due to weight and torque. Suspension can face higher loads. Cooling systems still need attention. Software can create new forms of service work.

Safety is a system property, not a slogan. EVs eliminate gasoline storage and combustion heat. That can reduce certain risks. EVs introduce high-voltage systems and battery fire risks in rare failure cases. Battery incidents can be serious when they occur. The correct approach is calm discipline. Follow manufacturer guidance. Use quality charging equipment. Avoid physical damage to the pack. Treat warning messages as real signals. In fleet settings, train staff on safe procedures.

Trade-offs are the heart of EV understanding. A larger battery can increase range. It can also increase weight and cost. More weight can reduce efficiency. More weight can also affect braking and tire wear. A more powerful motor can improve acceleration. It can also increase cost and heat management needs. Faster charging capability can improve travel convenience. It can also add cost and thermal complexity. There is no perfect design. There is only fitness for a job.

Many myths repeat in EV discussions. One myth is that "charging speed equals charger power." Charging speed depends on the vehicle and battery state. Another myth is that "range is fixed." Range changes with conditions and driving behavior. Another myth is that "EVs have no maintenance." EVs have different maintenance, not zero maintenance. Another myth is that "bigger battery always wins." Bigger battery can be wrong for cost and efficiency goals. Another myth is that "software fixes everything." Software can improve control and updates, but physics still sets limits.

Now consider a simple real example. Imagine a city commuter who drives 30 to 60 kilometers per day. That driver can often charge at home overnight. For this driver, daily convenience matters more than maximum range. A smaller battery may be enough. A smaller battery can also reduce cost and weight. The driver's key questions should be about home charging, reliability, and comfort in local weather. They should also ask about service support and warranty terms.

Now imagine a highway driver who regularly drives long distances. This driver's experience depends on fast charging access and station reliability. Range still matters, but charging curve and network uptime may matter more. The driver should ask about the vehicle's fast charging behavior across battery levels. They should ask how the car behaves in cold or very hot climates. They should also ask about route planning, charging payment friction, and backup options. The best vehicle for the commuter can be a poor match for the highway driver. The same model can feel "great" to one and "stressful" to another.

The key lesson is that an EV is not a single number. It is a linked system with constraints. If you learn the links, you stop arguing with slogans. You start asking the right questions for each use case. That skill will help you in every later chapter.

Quick Check

1. What is your primary use case: city commuting, mixed driving, or long highway travel?
2. Can you charge at home or work reliably most days?
3. Are you confusing energy (kWh) with power (kW) in your comparisons?
4. Which constraint matters most: cost, range, charging time, or reliability?

5. How will your climate affect heating, cooling, and battery behavior?
6. What is your acceptable risk level for new technology complexity?
7. What single factor would change your choice the most?

Chapter 2 — A Short History of EVs (What Changed, and Why)

Electric vehicles are not a new invention. Early electric vehicles existed long before today's market. In the early twentieth century, electric cars competed with steam and gasoline cars. They had a simple advantage. They were quiet and easy to operate. They did not require hand cranking and complex gear handling. They also had a major weakness. Batteries were heavy and limited in capacity. Charging infrastructure was limited as well. Gasoline systems improved rapidly and became widely supported.

The internal combustion ecosystem won for practical reasons. Fuel became easy to distribute at scale. Refueling became fast and familiar. Engines became more reliable and cheaper through mass production. Roads improved and travel distances increased. People wanted longer range and faster refueling. Electric technology could not match that with the batteries of the time. The result was a long period where EVs stayed niche.

The modern EV wave did not happen because people suddenly discovered electricity. It happened because several constraints changed at once. Battery technology improved over decades. Lithium-ion chemistry became widely used in consumer electronics first. That created supply chains, manufacturing learning, and safer designs. Over time, batteries gained better energy density and better management systems. They also gained better manufacturing quality and reliability.

A key shift was system integration. Early EV attempts often treated the battery as a simple component. Modern EVs treat the battery as a managed subsystem. The BMS became standard. Thermal management became standard. Power electronics improved in efficiency and control. Motors improved in materials and design. Software became central to performance and safety. These changes made EVs more practical in daily use.

Manufacturing scale also changed the story. Building batteries and EV components at scale changes learning rates. It reduces defects and improves consistency over time. It also attracts suppliers and service networks. This is not unique to EVs. It is how many industries mature. When volumes rise, quality systems become sharper. Costs can decline through improved yields and simpler assembly. However, costs can also rise temporarily when supply chains tighten. This is why the market can feel unstable during rapid expansion.

Policy and emissions pressure added another push. Many regions increased focus on local air quality and emissions reduction. These policies vary by country and time. Some policies support EV adoption with incentives. Some policies set fleet targets or emissions standards. The details change often, so this book stays high level. The stable point is that regulation can accelerate technology transitions. It can also shape which products survive. Compliance is not only a legal topic. It is a market force.

The modern wave also benefited from charging improvements. Home charging became a realistic option for many drivers. Public charging networks expanded in many regions. Standards and connectors developed, though differences remain. Reliability became a key issue for user trust. In many markets, charging experience became a competitive factor. This was a major shift from the early era. Early EVs lacked an ecosystem. Modern EVs are built into a growing ecosystem, even when it is imperfect.

Another trigger was the shift in user expectations. People became used to software-driven products. Phones and laptops normalized updates and feature improvements. EVs fit well into that trend because they rely on software deeply. Over-the-air updates can improve efficiency, user interface, and even safety features. This does not mean software can fix hardware limits. It means the product can improve without physical visits in some cases. That matters for customer satisfaction and long-term support.

Still, history also warns us. Technology waves can create mistakes. Rapid growth can expose quality gaps. Supply chain pressure can lead to variability. Software complexity can create new failures. Some EV launches faced issues related to manufacturing consistency, battery behavior, or charging compatibility. The details differ by company and model. The general lesson is stable. New technology at scale needs strong quality systems and honest communication. It also needs time for learning and correction.

Failures and recalls, when they happen, are not proof that the whole idea is broken. They are proof that systems require discipline. A recall can result from design choices, supplier issues, or manufacturing processes. In EVs, battery safety is one sensitive area. Another sensitive area is software validation. Another area is charging interoperability. A mature industry learns from these events. It improves standards, testing, and training.

The biggest change from earlier EV eras is the combination of batteries, software, and infrastructure. Older EVs were limited by batteries and lack of ecosystem support. Modern EVs still face battery limits, but the limits are higher. Modern EVs also have supporting networks and better control systems. This makes them practical for many jobs today. It also creates a path for further adoption as infrastructure improves.

History also explains why EV adoption looks uneven. Urban areas often adopt faster because driving patterns fit EV strengths. Fleets may adopt faster when routes are predictable and depot charging is possible. Long-haul and remote regions can be slower because infrastructure demands are harder. None of this is mysterious when you think in systems.

Quick Check

1. Which factor matters most in your region: battery cost, charging access, or regulation?
2. Are you treating EV adoption as a "car" issue or an "ecosystem" issue?
3. What part improved the most: batteries, software, or infrastructure?
4. What part still limits adoption: charging reliability, cost, or trust?

5. What does history suggest about early product issues in new waves?

6. Which segment adopts first: commuters, fleets, or long-haul users?

7. What would make adoption easier in your daily life?

Chapter 3 — EV Types and Use Categories (Not Just Body Styles)

Electric vehicles are often grouped by how they look. People say sedan, SUV, bus, or truck. That is useful for size, but it is not enough for decision clarity. EV systems behave differently across use cases. The right classification starts with power source and workflow. It then moves to duty cycle and environment. When you classify by job, you avoid expensive mismatches.

The first simple distinction is how the vehicle uses electricity and fuel. A battery electric vehicle, often called a BEV, runs only on electricity stored in a battery. It charges from the grid and does not have an engine for propulsion. A plug-in hybrid electric vehicle, often called a PHEV, uses a battery and also has an engine. It can drive in electric mode for some distance and then rely on fuel. A hybrid electric vehicle, often called an HEV, also has a battery and an engine. It usually does not plug in. It uses the engine and regenerative braking to manage energy.

These categories are not about "better" or "worse." They are about fit. A BEV can offer the simplest driving and the strongest electric benefits. It also depends most on charging access. A PHEV can reduce charging dependence in some situations. It can also add complexity because it carries two power systems. An HEV can improve efficiency without requiring a charger. It also does not remove fuel dependence. Understanding these trade-offs helps you choose rationally.

Now classify by vehicle class and job. Passenger cars are often used for mixed personal driving. Their duty cycle includes many short trips and occasional long trips. Buses and delivery fleets often run predictable routes. Predictable routes make charging planning easier. Trucks face heavy loads and long distances. Heavy loads require more energy. Long distances require more charging time or more stored energy. These constraints shape design choices strongly.

Two-wheelers and three-wheelers deserve separate attention. They are common in many markets and can electrify faster. They often need smaller batteries and simpler charging solutions. They can fit urban mobility and last-mile delivery well. Their economics can be different because fuel and maintenance savings may matter more. Their infrastructure needs can also be lighter.

Industrial and commercial EVs include forklifts, port vehicles, and specialized machines. These vehicles often operate in controlled sites. Controlled sites can support dedicated charging. That can accelerate adoption because infrastructure is easier to manage. The vehicle can also be optimized for one job, not many jobs.

The next classification is fleet versus consumer. Consumers often need flexibility. They may not plan routes carefully. They may not have guaranteed home charging. They also care about comfort and convenience. Fleets care about uptime, predictable cost, and operational planning. Fleets can build depot charging and manage schedules. Fleets also track data and maintenance systematically. This is why fleet adoption can look faster in some segments. It can also be slower if the business case is unclear or if capital budgets are constrained.

Urban versus long-haul is another critical split. Urban driving can favor EV efficiency because speeds are lower. Regenerative braking can recover energy during stop-and-go traffic. Home charging can cover daily needs. Long-haul driving puts pressure on charging networks and charging curves. High speed increases energy use. Cold weather can increase heating loads and reduce effective range. Long-haul EV design therefore often requires careful thermal and charging engineering.

A common mistake is matching the "right EV" to the "wrong job." People buy a long-range vehicle when they only drive short trips. They then carry extra battery weight every day. They also pay extra for a capability they rarely use. The opposite mistake is also common. People buy a small-battery vehicle for frequent highway travel. They then feel constant charging stress. They blame the whole technology, but the mismatch was the real issue.

Another mismatch happens with charging assumptions. Some buyers assume public charging will replace home charging easily. That can be true in some regions and false in others. Public charging reliability and access vary widely. The best EV for a home-charging driver may be a poor EV for an apartment driver without reliable charging. The vehicle is part of a larger system that includes living situation and infrastructure.

A third mismatch happens with climate. Cold climate drivers can face range reduction and slower charging without preparation. Hot climate drivers can face thermal management constraints under load. The same model can behave differently across climates. This is not unique to EVs, but it is more visible because batteries are sensitive to temperature. A good match accounts for the local climate and driving pattern.

You can also classify by performance and purpose. Some EVs are tuned for efficiency and cost. Some are tuned for high power and luxury. Some prioritize cargo and durability. Those choices affect battery size, cooling design, and pricing. They also affect maintenance and tire wear. The key is to pick the purpose first, then pick the vehicle.

At the industry level, these categories shape markets. Urban passenger BEVs can grow with home charging expansion. Fleet delivery BEVs can grow with depot charging and route optimization. Buses can grow where public procurement supports infrastructure. Trucks may grow in corridors where charging or other refueling solutions become reliable. The book will return to these themes later in markets and infrastructure chapters.

Quick Check

1. Which category fits your reality: BEV, PHEV, or HEV, and why?

2. Do you have reliable charging access where you live or work?

3. Is your driving mostly urban, mixed, or long-haul highway?

4. Are you buying for flexibility or for a predictable duty cycle?

5. What is your biggest constraint: cost, time, or charging access?

6. Which mismatch risk is highest for you: range stress or overpaying?

7. What local factor matters most: climate, roads, or charging network reliability?

Part 2 — The Technology Layer

Part Introduction: how EVs store, move, and manage energy

Part 1 gave you the map. Part 2 explains the engine under that map. Here, "engine" means energy storage, energy conversion, control, heat, and charging. These topics often get mixed together online. People argue about range without understanding temperature. People argue about charging speed without understanding charging curves. People argue about battery life without understanding degradation drivers.

This part keeps the story simple and connected. We start with the battery because it is the energy bank. We then move to the motor and inverter because they turn energy into motion. After that, we focus on heat. Heat is a hidden constraint that shapes performance, charging, and aging. We then explain charging in a practical way. Finally, we explain the software layer that makes modern EVs behave like managed systems, not just machines.

You do not need an engineering degree for this part. You only need patience and system thinking. The goal is not to memorize formulas. The goal is to understand trade-offs so you can think clearly.

Chapter 4 — Battery Basics for Non-Specialists

A battery in an EV is not one single block of energy. It is a structured system. It is built from many small energy units and managed by electronics. This structure exists for safety, performance, manufacturing, and service. When people say "the battery," they usually mean the entire pack. The pack contains cells, wiring, sensors, protection devices, and thermal hardware. It also contains the control logic that keeps everything inside safe limits.

A cell is the smallest energy unit in the pack. Think of a cell like a small container that can store and release electrical energy through chemical reactions. One cell is not enough for a car. A car needs high voltage and high capacity. Voltage helps deliver power efficiently. Capacity helps deliver energy over time. So many cells are connected together to reach the needed voltage and energy.

Cells are often grouped into modules. A module is a packaged group of cells. Not every design uses modules in the same way, but the idea is common. Modules can make assembly easier. Modules can also help with service and replacement in some designs. The full pack is the highest level. It is the complete battery unit installed in the vehicle. The pack includes the case, structural supports, electrical connections, sensors, and the cooling or heating pathways.

This structure also helps explain why "battery size" is not the whole story. A pack has a total capacity and a usable capacity. The usable capacity is often lower. That gap exists because safety margins are needed. The pack must avoid dangerous overcharge and deep discharge states. It must also allow control logic to balance cells and maintain stability over time. The exact margin depends on design choices. The stable point is simple. What you can use for driving is not always equal to the headline size.

Energy density is one of the most discussed battery ideas. It means how much energy is stored per unit of mass or volume. Higher energy density can help range. Higher energy density can also increase cost or create tighter safety constraints. This is why battery design is a constant trade-off. Engineers balance energy density, safety, cost, and lifespan. Consumers often want all four at the same time. Real products must choose priorities.

Safety in batteries is mostly about control and containment. A battery pack must handle normal driving and also handle abnormal events. Abnormal events can include a crash, internal defects, overheating, or charging faults. Safety systems include sensors that detect temperature and voltage behavior. Safety systems also include mechanical containment and electrical protection devices. Good safety design is layered. It assumes that one protection layer can fail. It therefore uses multiple layers.

Now consider degradation. Degradation is the slow loss of usable capacity and performance over time. It happens because the battery's internal chemistry changes with use and age. Some degradation happens even if you do not drive much. Time itself can cause changes. Some degradation is driven by cycling, which means charge and discharge. Some degradation is accelerated by heat and high stress. The key is to understand drivers rather than fear the idea.

Heat is one of the strongest drivers of battery stress. High temperature can accelerate chemical reactions that reduce capacity. Very low temperature can reduce short-term performance and charging acceptance. Temperature also affects internal resistance. When resistance rises, more energy is lost as heat. That heat then creates a feedback loop if not managed well. This is why thermal management matters so much. It is also why batteries behave differently in different climates.

Fast charging is another topic that needs calm clarity. Fast charging can be convenient. It can also increase stress in certain conditions. Stress is higher when the battery is cold, when it is very full, or when charging is pushed aggressively. Many modern EVs manage this by controlling charging power based on battery temperature and state of charge. This control is not a trick. It is a safety and longevity strategy.

State of charge, often written as SoC, is the system estimate of how full the battery is. It is an estimate, not a perfect measurement. The estimate can shift with temperature and aging. State of health, often written as SoH, is the system estimate of battery aging and remaining capacity compared to new. SoH is also an estimate. It can be measured and inferred in different ways. The stable point is that these indicators are designed for practical decision making, not laboratory perfection.

Another important idea is power versus energy. Energy is how much you can store and use over time. Power is how fast you can deliver that energy at a moment. A battery can have enough energy for long distance but still be limited in peak power under certain conditions. Temperature and internal design affect this. This helps explain why a vehicle may feel slower in extreme cold. It also helps explain why a vehicle may reduce peak performance after repeated hard use. The system protects itself.

Battery packs also face mechanical realities. They add weight. They change vehicle structure. They must be protected from impact and water intrusion. They must also be isolated electrically. High-voltage components require careful insulation and safe routing. These design decisions affect cost, service complexity, and crash performance. This is why batteries are not just "bigger equals better." Bigger changes the whole vehicle system.

If you keep one mental model from this chapter, keep this one. A battery is an energy bank managed by controls and protected by layers. Its behavior depends on temperature, use pattern, and design choices. This is why two EVs with similar battery size can feel different. It is also why real range and charging experience depend on more than the headline numbers.

Quick Check

1. Can you explain the difference between a cell, a module, and a pack in one sentence each?

2. Are you mixing "energy" with "power" when comparing vehicles or chargers?

3. Which factor dominates battery behavior in daily life: temperature, speed, or driving style?

4. Why does usable capacity often differ from total pack capacity?

5. What does SoC tell you, and what does it not tell you?

6. What does SoH try to represent, and why is it an estimate?

7. Which habit is likely to reduce stress: gentle temperature management or chasing maximum charge speed?

Chapter 5 — Motors, Inverters, and Power Electronics

The motor is the EV's main machine for motion. It converts electrical energy into mechanical rotation. In a conventional car, the engine produces power through combustion and heat. In an EV, the motor produces power through electromagnetic forces. The result is a different driving feel. Torque can be delivered quickly. Control can be precise. The vehicle can respond smoothly to small changes in the accelerator.

A traction motor is designed to move the vehicle. It must operate across a wide speed range. It must also handle different loads, from gentle cruising to steep climbs. It must be efficient enough to protect range. It must also be robust enough to survive vibration, heat, and repeated high-power use. This is not a simple design problem. It is a balance problem.

The inverter is the motor's partner. The battery stores energy as direct current. Many traction motors require controlled alternating current to operate efficiently. The inverter converts DC from the battery into AC for the motor. It also controls frequency and voltage patterns to manage motor speed and torque. In simple terms, the inverter is the "power steering wheel" for electricity going into the motor.

This is why the inverter affects both performance and efficiency. Better control can reduce losses. Better control can also reduce heat. Heat matters because power electronics can be sensitive to high temperature. These components often use specialized materials and careful cooling. This adds cost and complexity, but it improves reliability and performance.

Efficiency is not a single number you carry forever. Efficiency changes with speed, load, and temperature. At low speeds with frequent stops, regenerative braking can help recover energy. At high speeds, aerodynamic drag rises strongly, so energy use increases. Climbing hills requires more power. Descending hills can recover some energy, but not all. Cold weather can add heating load and change battery behavior. So real efficiency is a pattern, not a constant.

Regenerative braking is one of the most valuable motor features. When you slow down, the motor can act like a generator. It converts motion back into electrical energy and sends it into the battery. This can improve efficiency, especially in city driving. It also changes driving style. Some vehicles offer strong one-pedal driving. That means lifting your foot slows the car with regeneration. The system still uses friction brakes for stronger stops and safety cases.

Power electronics also includes converters. Most EVs still have a low-voltage system for lights and electronics. The EV must supply that system from the main battery. A DC-to-DC converter usually does this. The vehicle may also manage different voltage domains for different subsystems. The goal is stable power delivery and safe isolation between systems.

Common failure points exist, but they are better understood as "risk areas" rather than guaranteed problems. Power electronics can be stressed by heat. They can also be stressed by repeated high load and rapid cycling. Motors can face bearing wear and insulation stress over very long use. Cooling systems can degrade if not maintained. Software calibration errors can create unpleasant behavior or reduced efficiency. The stable point is that EV drivetrains can be durable, but they still need disciplined design and testing.

Another important idea is torque delivery and traction. EV torque can arrive quickly. That is great for responsiveness. It can also increase tire wear and traction challenges if not managed well. Control software limits torque based on grip conditions and stability goals. This is why the control layer is deeply linked to the motor layer. The hardware can deliver power, but software decides when and how much.

EV drivetrains often use simpler mechanical gear setups than combustion cars. Many use a single-speed reduction gear. This reduces mechanical complexity. It can also reduce some maintenance types. It does not remove all mechanical needs. Bearings, joints, and gears still exist. The point is to recognize where complexity shifted. Some complexity moved from mechanical systems to electronics and software.

When you read a performance claim, ask the right questions. Is it peak power or sustained power? Is it measured in ideal temperature conditions? Does the vehicle limit power after repeated high use to protect heat limits? Does the vehicle behavior change at low state of charge? These questions are not "negative." They are realistic. They help you understand how the system protects itself and how it will behave in your use case.

If you keep one mental model from this chapter, keep this one. The battery provides energy. The inverter shapes that energy into controlled power. The motor turns power into motion. Efficiency and heat depend on how well these pieces work together, and on the driving conditions.

Quick Check

1. Can you explain the motor's job in one sentence without using jargon?
2. Why does the vehicle need an inverter if it already has a battery?
3. What is the difference between energy recovery and friction braking?
4. Why is efficiency a pattern rather than a constant number?
5. What makes high-speed driving harder on range than city driving?
6. Which risk area is most linked to reliability: heat, software, or mechanical wear?
7. What does "sustained power" mean in your own words?

Chapter 6 — Thermal Management and Safety

Heat is often the hidden constraint in EVs. People focus on battery size and motor power. Heat quietly sets the real limits. Heat shapes charging speed. Heat shapes repeated performance. Heat shapes battery aging. Heat also shapes safety, because uncontrolled heat can lead to serious failure modes.

EVs create heat in several places. The battery creates heat when it charges and discharges. The inverter creates heat during conversion. The motor creates heat under load. Even the cabin heating system can affect energy use and battery temperature strategy. A good EV does not only "cool things down." It manages heat actively and intelligently.

Thermal management means keeping components inside a safe operating window. That window depends on the component. Batteries prefer a moderate range for best performance and longevity. Power electronics also prefer controlled temperature. Motors can handle heat, but they still have limits. If temperature goes outside the target window, the system reduces performance or charging speed. This behavior is not a defect by itself. It is often protection.

Cooling strategies can vary. Some vehicles use liquid cooling for the battery pack. Some use air-based methods in certain designs. Liquid cooling can provide more stable control, but it adds complexity. It needs pumps, lines, and heat exchangers. It also needs quality sealing and maintenance planning. The details depend on design goals and cost targets. The stable idea is that better thermal control often enables better performance and charging consistency.

Cold weather shows the other side of the thermal story. When the battery is cold, internal resistance increases. Charging acceptance often reduces. The system may need to warm the pack for fast charging. Cabin heating can also draw energy. This can reduce effective range, especially on short trips where the battery and cabin start cold. Planning matters here. Preconditioning can help in some systems. That means warming or cooling the battery before driving or charging.

Thermal runaway is a term that is often misused. Thermal runaway is a dangerous condition where heat inside a cell increases rapidly and feeds itself. It can be caused by internal defects, severe damage, or extreme overheating. Most daily EV use does not bring a battery near this state. Safety design aims to prevent the conditions that can trigger it, and to contain it if it happens. It is not helpful to treat every EV as a fire risk. It is also not helpful to deny that battery incidents can be serious. Calm realism is the correct stance.

Safety design mindset in EVs is layered. The system tries to prevent failures. It tries to detect abnormalities early. It tries to isolate faults. It tries to limit escalation. It also tries to protect occupants in crashes. Battery packs are often placed low for stability, but that also requires strong structural protection. High-voltage components require clear isolation strategies. Service procedures require training. All of this is part of responsible design.

Now consider a real example that many drivers notice. A driver in a hot climate may see reduced charging speed on a very hot day, especially after highway driving. The pack is already warm from discharge. Fast charging adds more heat. The system may reduce charging power to keep temperature inside limits. The driver may interpret this as "the charger is slow." The true reason is often thermal protection.

A driver in a cold climate may see slow fast charging early in a trip. The pack may start cold. The system may limit charging power until temperature rises. The driver may interpret this as "the car cannot fast charge." The true reason is often temperature. In both cases, the experience depends on the whole system, not just the charger rating.

The practical takeaway is simple. Heat management is part of daily EV reality. Good planning includes understanding climate and route. It includes giving the system time to warm or cool when needed. It also includes safe charging habits and avoiding physical damage. Thermal management is not glamour. It is what makes the rest of the technology usable and durable.

Quick Check

1. Why does heat shape both charging speed and battery aging?

2. What is the difference between "warm battery" and "unsafe battery"?

3. Why can cold weather reduce charging acceptance?

4. What does "preconditioning" mean in this context?

5. Why might fast charging slow down after highway driving in hot weather?

6. What is thermal runaway, and why is it not a daily-driving event?

7. What is one safe habit you can apply regardless of vehicle brand?

Chapter 7 — Charging Hardware and Charging Speed Reality

Charging is the EV refueling system. It is also the most misunderstood part of the ecosystem. People often talk about charging speed as if it is a single fixed number. In reality, charging is a conversation between the charger and the vehicle. The charger offers power. The vehicle decides how much it can accept safely. That decision changes with battery temperature and state of charge.

AC charging uses alternating current, which is common in homes and many workplaces. The vehicle's on-board charger converts AC to DC for the battery. AC charging is usually slower than fast charging. It is often enough for overnight charging. For many drivers, AC charging is the most important type because it supports daily routine without stress.

DC fast charging provides direct current to the vehicle. It can charge the battery faster, especially when the battery is at a lower state of charge and at a suitable temperature. DC fast charging is more complex. It depends on charger capability, grid connection, station health, cable cooling, and communication standards. It also depends heavily on the vehicle's pack design and cooling strategy.

Charging curves explain why "max kW" is not the full story. Many EVs can accept high power at low state of charge. As the battery fills, the system reduces power. This is normal. It protects the battery and prevents overheating. It also helps the battery reach a balanced state. So the best metric for road trips is often not peak power. It is how quickly the vehicle adds usable range across a typical charging window.

Connectors and standards matter because they shape real access. Different regions and networks can use different connector types and protocols. This book stays high level because details change. The stable point is that interoperability affects daily convenience. The more compatible and reliable the ecosystem, the easier EV adoption feels. When compatibility is fragmented, user friction increases.

Home charging and public charging also differ in economics. Home charging can be cheaper and more convenient in many settings. It can also be limited by home electrical capacity and installation constraints. Public charging can be essential for apartment living or long travel. It can also be more expensive and less predictable. Pricing models vary. Availability varies. Reliability varies. So "best choice" depends on living situation, local infrastructure, and driving pattern.

If you keep one mental model from this chapter, keep this one. Charging speed is not a promise. It is a result of battery state, temperature, and station condition. The best planning focuses on routine charging first, and fast charging second.

Quick Check

1. What is the main difference between AC charging and DC fast charging?
2. Why does the vehicle control how much power it accepts?

3. What is a charging curve, and why does power taper?
4. Which matters more for daily life: home charging access or peak fast-charge power?
5. Which matters more for road trips: peak power or time to add usable energy?
6. What local factors shape charging experience: price, reliability, or coverage?
7. How would your conclusion change if you could charge at home every night?

Chapter 8 — Software, Controls, and the EV "Digital Layer"

Modern EVs behave like managed systems because software is deeply involved. Software controls energy flow, torque delivery, charging behavior, thermal strategy, and safety limits. This does not mean "software replaces physics." It means software helps the vehicle operate close to its best balance across different conditions.

The battery management system, or BMS, is one of the most important control layers. The BMS measures cell voltages and temperatures. It estimates state of charge and state of health. It also manages cell balancing. Cell balancing helps keep cells aligned so one weak cell does not limit the pack early. The BMS also enforces safety limits. It can limit charging or power output if conditions are risky. This can feel inconvenient in a moment. It can also protect the system over years.

Energy estimation and range prediction are also software-heavy. Range estimates are not fixed truth. They are predictions based on past driving, current conditions, and system models. Weather, speed, elevation, tire pressure, and heating use can change outcomes. So range prediction has limits. A good driver understands that the estimate is guidance, not certainty. This mental stance reduces stress and improves planning.

Over-the-air updates are another software feature. OTA updates can fix bugs, improve user interface behavior, and adjust energy management. They can also introduce risk if update processes are not disciplined. This is why update strategy is part of trust. A well-managed update process includes testing, staged rollout, and rollback plans. The details depend on the company. The stable point is that software updates are a powerful tool, but they must be handled responsibly.

Driver assistance and autonomy are also often mixed with EV discussions. They are separate topics but they share software foundations. Driver assistance can include lane keeping, adaptive cruise control, and automatic emergency braking. Autonomy claims should be treated carefully. Boundaries depend on sensors, environment, and safety validation. This book stays high level. The stable idea is that driver assistance supports the driver. It does not remove the driver's responsibility in most real-world settings.

If you keep one mental model from this chapter, keep this one. Software is the coordinator of the EV system. It can improve usability and safety. It also adds complexity that must be managed with quality discipline.

Quick Check

1. What is the BMS trying to protect: safety, lifespan, or both?
2. Why is range prediction a model, not a promise?
3. What daily factors can change range more than people expect?
4. What benefits can OTA updates provide in a mature system?
5. What risks can OTA updates introduce if quality discipline is weak?
6. Why should autonomy claims be treated cautiously?

7. What is one way software and thermal management are linked?

Part 3 — Charging Networks and the Grid

Part 3 moves outside the car. It looks at charging networks and the grid that supports them. Many EV frustrations are not inside the battery or the motor. They are about access, reliability, and planning.

You will see how charging sites are built and operated at a basic level. You will also learn what "good coverage" means in real life. It is not only the number of chargers. It is location, uptime, and simple payment.

Then we step back to the grid. We explain peak demand, managed charging, and why timing matters. We also cover V2H and V2G in a calm way. We focus on where they help first, and why adoption is often slower than headlines.

Chapter 9 — Charging Networks as Infrastructure

Charging networks look simple from a driver's seat. You plug in, you wait, and you leave. Behind that simple action sits a full infrastructure business. It is closer to telecom and utilities than to retail. A charging site must deliver power safely. It must communicate with the vehicle. It must verify payment. It must record sessions. It must protect users and equipment. It must also work in heat, dust, rain, and heavy daily use.

A charging network is not only "chargers on a map." It is a system of hardware, software, and operations. Hardware includes the charging units, cables, connectors, and cooling systems. It also includes transformers, switchgear, protection devices, and site wiring. Software includes the station controller, the user app, the payment and identity layer, and monitoring tools. Operations include field technicians, spare parts, remote support, and service contracts. Without operations, hardware becomes broken hardware.

This is why the word "infrastructure" matters. Infrastructure is not judged by marketing. It is judged by uptime and trust. Drivers do not remember a charger that worked once. Drivers remember the network that works when they need it. Businesses think in the same way. A fleet manager cannot accept "maybe it works." They need predictable daily charging. Reliability becomes the product.

Now consider how networks are built. A site begins with a location decision. The location needs power access, parking layout, and safe traffic flow. It also needs demand. Demand can come from commuters, travelers, or fleets. The site then needs permission from the property owner. It often needs local approvals and safety inspection. It also needs a grid connection that can support the power level. This grid step can be easy or hard depending on location. It can require upgrades in some cases.

After approvals, equipment must be installed. Installation is not only placing a charging unit. It includes trenching, conduit, wiring, protective devices, grounding, and signage. It includes testing and commissioning. Commissioning is the step where the site is proven to work as a system. It must deliver power safely. It must stop when needed. It must communicate correctly. It must report status correctly. If commissioning is weak, failures appear later in real use.

Maintenance is the part drivers rarely see. Cables wear out because they are handled many times. Connectors are exposed to stress, dirt, and heat. Screens can fail from weather and vandalism. Cooling systems can clog or leak. Payment terminals can fail. Sensors can drift. Network software can also break from updates or communication faults. A good network has monitoring that detects problems early. It also has technicians who fix issues quickly. It also has spare parts ready. Without this, small failures turn into long outages.

Reliability and uptime are therefore not a "nice extra." They are the core of user experience. A driver plans time and route. If a charger fails, the driver loses time and feels risk. If the driver arrives with low battery, the risk feels higher. Even if the driver is safe, the stress is real. This stress becomes a barrier to adoption. People may avoid EVs, not because EVs are bad, but because infrastructure trust feels weak.

Network reliability also shapes charging behavior. If drivers trust a corridor network, they can charge in shorter sessions and keep moving. If drivers do not trust the network, they may "overcharge" when they find a working station. Overcharging here means charging far above what is needed for the next stop. This behavior increases congestion. It also increases waiting time for others. It also increases frustration at busy sites. Poor reliability can therefore create a negative cycle.

Payment and access are another major source of friction. Drivers want simple, consistent access. In practice, networks can have different apps and different membership rules. Some sites accept contactless payment easily. Some sites require an app and account setup. Some sites support roaming, which means a user can use one account across multiple networks. Roaming can reduce friction, but it is complex. It needs agreements, data exchange, and settlement processes. If roaming is partial or inconsistent, it can confuse users.

Pricing can also be confusing. Pricing can include energy-based pricing, time-based pricing, or mixed pricing. The details can depend on local rules and network policy. Drivers care less about the pricing model and more about clarity. They want to know the real cost before they start. They want billing to match what they used. They also want customer support if something goes wrong. This is a trust topic again, not only a money topic.

"What good coverage really means" is often misunderstood. Many people think coverage means "many pins on the map." Pins do not guarantee working charging. Good coverage means you can complete typical trips with low stress. It means you have backup options when one station is down. It means stations are placed where people actually stop. It means power levels match the use case. It also means congestion is managed, especially at peak travel times.

Coverage has layers. Home and workplace charging cover routine life. Public fast charging covers travel and urgent needs. Depot charging covers fleets. Street charging can support people without private parking. Each layer serves a different need. A network that focuses only on one layer may leave gaps for many users. A complete ecosystem needs coordination between these layers.

A practical way to think about a network is to imagine it as a "service promise." The promise is not a specific speed every time. The promise is a reliable path to energy. For travel corridors, that means stations spaced so drivers have options. For city life, it means chargers where people park long enough. For fleets, it means high uptime and predictable schedules. The best networks design for these realities.

Here is a real example with simple details. A driver lives in an apartment without home charging. The driver relies on public chargers near work and shopping. If chargers are often broken or blocked, the driver must spend time searching. This time becomes a hidden cost. The driver may shift their routine around charging. That feels like a burden. If the city has reliable curbside chargers and a reliable fast charging hub, the driver's life becomes easier. They can charge while doing normal tasks. The EV becomes practical.

Now consider a highway travel example. A family travels on weekends. They need fast charging on the route. If there are only one or two stations and one is down, the trip becomes uncertain. The family may avoid EV travel. If there are several reliable sites and real-time status is accurate, travel becomes normal. The family can plan a short stop. They can choose a backup. The trip becomes calm.

Common mistakes happen at three levels. Drivers make mistakes, networks make mistakes, and public planners make mistakes. Drivers may focus only on peak charging speed and ignore reliability. Drivers may arrive at busy times without a plan. Drivers may block chargers after charging ends. Networks may expand too fast without enough maintenance capacity. Networks may install hardware without local climate fit. Networks may also push updates without careful testing. Planners may approve sites that look good on paper but are hard to access. Planners may also ignore signage and traffic flow, which creates unsafe layouts.

The discipline here is the same discipline used in quality systems. Define the user journey. Identify failure points. Monitor performance. Fix root causes. Repeat. Charging networks are a service business with physical assets. They need operational excellence as much as they need capital.

Quick Check

1. What are the three layers of a charging network system beyond the charger itself?

2. Why can "many pins on a map" still produce a bad user experience?

3. Why does maintenance capacity matter as much as new site expansion?

4. What types of friction can payment and access create for drivers?

5. How can poor reliability increase congestion at working chargers?

6. What does "good coverage" mean in practical travel terms?

7. Which layer matters most for your daily life: home, workplace, or public?

Chapter 10 — Grid Impact and Smart Charging (High Level)

The electric grid is the delivery system for all charging. Most drivers only meet the grid through a wall outlet or a charging station. Behind that outlet sits a large balancing act. Electricity must be produced and delivered at the same time it is used. Large-scale storage exists in some places, but the system still relies on balance. When many EVs charge together, the grid sees a new pattern of demand. This is not automatically a crisis. It is a planning and management challenge.

The simplest concept is peak demand. Peak demand is the highest load that happens during a period. It is the moment when many users consume electricity at once. Peak demand matters because equipment must handle it safely. Transformers, feeders, and substations are sized for expected peaks. If new loads push peaks higher, upgrades may be needed. Upgrades cost money and time. Utilities therefore care about when charging happens, not only how much charging happens.

Load balancing is the idea of shaping demand to avoid sharp peaks. If charging is spread across time, the grid can serve more energy with less stress. This is the core logic behind managed charging. Managed charging means charging is controlled or scheduled to match grid conditions and user needs. Control can be simple or advanced. It can be done by the car, the charger, the building, or the utility, depending on design and rules.

Time-of-use pricing is one method to influence charging time. In simple terms, electricity can cost more during high demand periods. It can cost less during low demand periods. This encourages users to shift flexible loads. EV charging is often flexible for many users. Many vehicles are parked for long periods overnight. If drivers charge during those periods, peaks can be reduced. If drivers charge during evening peaks, peaks can rise. Price signals can therefore shape behavior without strict control.

Managed charging is not only about saving money. It is also about reliability. A building with many EVs can overload internal electrical capacity if charging is unmanaged. The building may need upgrades. Or it can use smart load sharing. Load sharing means chargers communicate and limit total power to stay within a safe cap. Each vehicle may charge slower, but all vehicles can charge enough for daily needs. This approach can reduce upgrade needs and improve overall satisfaction.

Workplace charging has its own patterns. People arrive in the morning and park for hours. Slow or moderate charging often fits well. If every workplace installs high-power chargers without management, it can create a midday load spike. If workplaces use smart scheduling and load caps, they can support many vehicles with modest capacity. The key is matching charging power to parking duration. Parking time is an asset. It reduces the need for extreme power.

Depot charging for fleets is another special case. Fleets often return at similar times. They may need vehicles ready by morning. If all vehicles charge at once at full power, depot demand can spike. Depots can manage this through staged charging. They can prioritize vehicles based on next-day routes. They can also charge more during lower demand hours. This is an operations problem as much as an electrical problem. The best solutions connect energy planning to fleet scheduling.

Utilities care about predictable load and safe limits. They care about transformer loading and feeder congestion. They care about local voltage quality and stability. They care about outage recovery and safety. They also care about long-term planning. If EV adoption rises, utilities want visibility. They want to know where load will grow. They want to invest ahead of time, not after failures. This is why data and coordination matter.

A clear mental model helps here. Think of the grid as a road system. If everyone drives at the same hour on the same road, traffic jams happen. If travel is spread, the same road can serve more total trips. Managed charging is like traffic management. It aims to spread load across time and reduce congestion at peak moments.

Now consider a practical neighborhood example. A suburban area has many homes with evening activity. Cooking, heating or cooling, and lighting create an evening peak. If many EVs start charging as soon as people arrive home, that peak can rise sharply. The local transformer may run hot. The feeder may approach limits. If instead most EVs begin charging later at night, the peak can be lower. The energy delivered overnight can be the same. The difference is timing. Timing can reduce stress and reduce upgrade needs.

Another example is a public fast charging hub near a highway. Fast chargers can draw significant power during busy travel times. That site may need dedicated electrical upgrades. It may need on-site equipment that manages peak draw. It may also need coordination with the utility. If the site is designed well, it can serve many travelers reliably. If the site is designed poorly, it can suffer from power limits and outages. Again, the grid link is part of user experience.

Smart charging can also improve resilience. In some designs, charging can pause during a local constraint. It can resume when conditions improve. This can prevent overload. It can also protect building systems. Drivers may worry that "smart" means losing control. In good systems, drivers still set goals. They set a target state of charge and a departure time. The system optimizes within those preferences. The driver gets what they need with less grid stress.

There are also valid concerns. Managed charging needs trust and privacy discipline. Users want transparency about what is controlled. They want clear opt-in rules in many contexts. They want safe and fair pricing. They also want reliability. A system that delays charging but fails to meet morning needs will not be accepted. So smart charging must be designed around outcomes, not around control for its own sake.

The key point is balanced thinking. The grid is not an enemy of EVs. The grid is the enabling system. EV charging adds demand, but it also adds flexibility. That flexibility is valuable if managed well. It can support better utilization of existing assets. It can also support future energy transitions, depending on local energy mix. Those details vary by place and policy, so this book stays high level and focuses on the stable principles.

Common mistakes include treating every charger as a high-power device. Many daily use cases do not need high power. Another mistake is ignoring building electrical limits until problems occur. Another mistake is installing chargers without load management in multi-unit housing. Another mistake is using price signals without clear communication. People resist what they do not understand. Clarity and simple rules improve adoption.

Quick Check

1. What is peak demand, and why do utilities care about it?

2. What is load balancing in simple language?

3. Why can slow charging be "better" when parking time is long?

4. What is managed charging, and who can control it?

5. How can time-of-use pricing change charging behavior?

6. Why do fleet depots need charging plans tied to schedules?

7. What is one way smart charging can improve reliability?

Chapter 11 — Vehicle-to-Home and Vehicle-to-Grid (V2H/V2G)

Bidirectional charging is one of the most exciting ideas around EVs. It is also one of the most misunderstood. The basic idea is simple. Electricity can flow in two directions. The EV can take energy from the grid to charge. The EV can also send energy back out under controlled conditions. When energy flows to a home, it is often called vehicle-to-home. When it flows to the grid, it is often called vehicle-to-grid. Some discussions also include vehicle-to-load, which means powering devices directly. The names vary, but the core idea is energy export.

Bidirectional charging sounds like a perfect solution. You have a large battery in your driveway. You could use it as backup power. You could support the grid during peaks. You could potentially earn value by providing services. These are real possibilities in some contexts. They are not universal. They require hardware support, software coordination, and regulatory permission. They also require clear economics. Without those conditions, adoption stays slower than headlines.

Start with what bidirectional charging actually means. The EV and the charger must be designed for two-way power flow. A normal charger may only deliver power one way. A bidirectional system needs conversion equipment that can send AC power to a home or to the grid in a stable way. It also needs safety controls. The system must isolate from the grid during outages to protect line workers. This is sometimes called anti-islanding protection. Details vary by region and rules, so we stay conceptual here. The key is that exporting power requires strict safety design.

Vehicle-to-home use is often the easiest first step in many settings. A home can use the EV as backup during an outage. The EV battery can power essential loads. Essential loads can include lighting, refrigeration, and basic devices, depending on setup. The value here is resilience. The user values having power when the grid is down. This is not about profit. It is about safety, comfort, and continuity.

Vehicle-to-home can also help with energy shifting. A home might charge the EV during low-cost hours and use some energy during high-cost hours. This depends on pricing structure and permission. It also depends on how the user values battery usage. The system must protect the vehicle's main job, which is driving. Most users will not accept losing mobility for small savings. A good system therefore keeps a reserve for driving.

Vehicle-to-grid is more complex. Exporting to the grid is not only a technical matter. It is also a market and regulation matter. Grid operators care about stability and predictability. They need resources that can respond in controlled ways. They also need measurement and verification. They need to know how much power is delivered and when. They need confidence that resources will show up when called. A single vehicle is not reliable enough on its own. Many programs therefore aggregate many vehicles. Aggregation means a platform manages a fleet of participating EVs and coordinates export.

Where does V2G make sense first? It tends to make sense where vehicles are parked reliably for long periods and can be managed. Fleet vehicles can fit this in some cases. Workplace parking can fit in some cases. Public sector fleets can fit because schedules are stable. School bus fleets are sometimes discussed because they have long idle periods, but real suitability depends on local conditions and vehicle design. The stable principle is this. V2G needs predictable parking patterns and predictable participation.

Battery wear is a common worry, and it should be treated with balance. Any use of the battery involves cycling, and cycling contributes to aging. The amount of added wear depends on depth of discharge, temperature, and current levels. Small controlled cycles might add modest wear. Large frequent cycles can add more wear. The true impact varies by design and usage. So it is not honest to promise "no wear." It is also not honest to assume "it destroys the battery." The right approach is cautious and transparent. Programs need clear rules and incentives that reflect battery cost and warranty terms.

Warranty concerns are also real. Warranties vary by manufacturer and by region. Some warranties may allow certain forms of export. Some may not. Some may allow only approved equipment. Because these terms can change, and because they differ, this book treats them as "unknown until you check your specific case." The stable guidance is simple. Do not assume your vehicle supports export. Do not assume it is permitted. Verify support and terms before you invest.

Another reason adoption is slower is hardware complexity and installation cost. Bidirectional home systems can require additional equipment and electrical work. They may also require permits and inspections. They may require special switchgear. They may require specific charger models. All of this adds friction. A technology can be impressive and still be slow to spread because installation is not simple.

There is also a human factor. Many people want charging to be effortless. Adding export control adds choices and settings. It adds "what if" worries. What if I forget to keep enough charge for a trip? What if the system exports too much? What if it fails during an outage? Good designs reduce these worries through simple defaults. They maintain a driving reserve. They provide clear status. They allow easy override. Without these, adoption stays niche.

Let's use a realistic example for V2H. A family lives in an area with occasional outages. They own an EV and have home solar in some cases, but solar alone does not help at night without storage. A bidirectional setup allows the EV to power essential loads during outages. The family sets a rule. The vehicle never goes below a reserve that supports their daily commute. During an outage, the system powers the refrigerator, basic lighting, and phone charging. This improves resilience. It does not require interaction with grid markets. It simply provides peace of mind.

Now a realistic example for V2G. A delivery fleet returns to a depot each evening. Vehicles are parked overnight. The depot has a managed charging system. The grid sees a peak in the evening. A local program offers incentives for reducing peak stress. The depot can reduce charging during the peak. It can also export a small amount of energy if permitted and compensated. The key value is predictable participation. The fleet has predictable parking. The program has predictable response. This is closer to how V2G can scale: through managed fleets and aggregated systems.

There is also a broader system benefit. If V2G scales in controlled ways, it can act as a flexible resource. It can help smooth peaks. It can help integrate variable renewable generation in some contexts. These system claims depend on local grid design and policy, so we keep them high level. The stable principle remains: flexible load and flexible export can support stability if coordinated well.

Common mistakes in this topic are driven by hype. One mistake is assuming every EV can do bidirectional export. Another mistake is assuming export is always profitable. Another mistake is ignoring installation complexity and safety requirements. Another mistake is ignoring the main purpose of a vehicle, which is reliable transport. A good system must protect mobility first. It must also protect safety first. It must also respect local rules.

The best mindset is practical. Bidirectional charging is a tool. It will grow where it fits. It will grow first in resilience use cases and structured fleet programs. It will grow slower in casual consumer use unless installation becomes simpler and rules become clearer. This is not failure. It is normal technology adoption behavior.

Quick Check

1. In one sentence, what does bidirectional charging mean?

2. What is the difference between V2H and V2G in purpose?

3. Why does exporting power require strict safety controls?

4. Why do aggregated programs matter for V2G reliability?

5. What factors influence battery wear in export scenarios?

6. Why can adoption be slow even if the idea is good?

7. What rule would protect mobility in a home backup setup?

Part 4 — Design, Manufacturing, and Quality Systems

An EV can feel simple when you drive it. You press the pedal and it moves. You plug in and it charges. But building EVs at scale is not simple. It is a large system with many trade-offs. It includes design choices, factory processes, software discipline, and quality control. It also includes supply risks and cost pressure. This part explains how EVs are designed and built in a practical way. It focuses on what stays true across brands. It avoids brand hype and "magic technology" claims. It treats quality as a system, not a slogan. It also treats safety as a design requirement, not an afterthought.

Chapter 12 — EV Platform Design and Packaging

A vehicle platform is the basic architecture that holds the whole car together. It is the structure that sets where major parts go. It sets how weight is carried. It sets how space is used. It also sets which designs are easy, and which are expensive. In EVs, the platform also shapes the battery pack location. It shapes the motor location. It shapes crash structures. It shapes cooling routes. It shapes wiring and electronics. When a platform is chosen, many later decisions become easier or harder.

People often hear the phrase "skateboard platform." The simple idea is this. The battery pack sits low in the floor area. The main power electronics and motors fit around it. The wheels and suspension attach to a strong frame. The body sits on top like a cabin. The name is only a picture. The real point is packaging. Packaging means fitting everything into a limited space. Packaging also means keeping the vehicle safe, serviceable, and manufacturable.

A skateboard-style layout can give EVs strong benefits. A low battery pack lowers the center of gravity. That can improve stability. It can also reduce body roll. It can improve handling feel. It can also free space in the front. Some EVs use that space for storage. Some use it for thermal and power electronics. The platform can also be modular. Modularity means the same base can support many car shapes. That can reduce cost. It can also speed up new models.

But every benefit comes with trade-offs. Placing a large battery pack in the floor changes the vehicle's structure. The floor must be strong. It must protect the pack in crashes. It must also protect it from road impacts. The pack housing adds weight. That weight competes with range and cost. The pack also needs cooling and heating paths. Those paths need space and complexity. The battery also affects how the cabin floor is shaped. It can change seat height. It can change driving position. These are real packaging decisions.

Space trade-offs also appear in cabin design. A higher floor can reduce foot room. Designers may raise the seat. That can change comfort. In small cars, this is harder. In larger SUVs, it can be easier. Another trade-off is ground clearance. If the pack is low, it can be vulnerable. If the pack is high, the center of gravity rises. Designers choose a balance based on vehicle purpose and market needs.

Weight budgeting is another core discipline. Weight is not "one number." It is a set of choices across the car. Batteries add weight compared to a fuel tank. Motors and power electronics add weight too. But EVs can also remove weight. They remove the engine and many mechanical parts. The total outcome depends on design. Weight budgeting means allocating weight targets to each major subsystem. It is like budgeting money. If one system gets heavier, another system must get lighter, or the whole vehicle becomes heavier.

Weight matters because it affects energy use. It affects braking. It affects tire wear. It affects crash forces. It also affects cost. In EVs, weight matters strongly because the battery must carry the car's mass. A heavier car needs more energy for the same trip. That can reduce real range. Designers respond by using stronger but lighter materials. They also respond by improving aerodynamics. They also optimize rolling resistance. But these improvements have limits.

Structural design also changes in EVs. The platform must handle torsional stiffness. That is the platform's resistance to twisting. Stiffness matters for handling and noise control. It also matters for durability. A stiff structure can reduce rattles over time. But stiffness can add material and weight. The platform must also manage crash loads. Crash energy must be absorbed and redirected. It must not crush the battery pack. It must protect passengers. It must also protect high-voltage systems from dangerous failure.

Crash safety in EVs includes normal vehicle crash concepts. It also adds battery safety concepts. The battery pack is a large energy store. It must remain stable during a crash. The pack enclosure must resist intrusion. The pack must be isolated electrically if needed. Many designs include sensors that detect crash events. They can command contactors to open. Contactors are controlled switches that can disconnect the high-voltage pack from the rest of the vehicle. This is one layer of protection. There are also fuses and isolation monitoring. The details vary by design, but the safety goal stays consistent. Keep people safe, and keep electricity under control.

Crash safety also affects service. After a crash, a battery may look fine outside, but it may have internal damage. Internal damage can create delayed risk. That is why post-crash inspection rules matter. These rules differ by region and manufacturer. This book stays high level. The key idea is that energy systems need careful inspection after impacts. A calm, cautious approach is appropriate.

NVH is another important area. NVH means noise, vibration, and harshness. In a fuel car, the engine creates noise and vibration. That noise can hide other sounds. In an EV, the engine noise is removed. That sounds like a benefit, and it is a benefit. But it also reveals other noises. Tire noise becomes more noticeable. Wind noise becomes more noticeable. Gear whine from the motor reduction gear can become noticeable. Cooling fans can become noticeable. Even small rattles can become noticeable.

EV vibration behavior can also differ. Electric motors deliver torque quickly. That can stress mounts and joints in different ways. Regenerative braking can also change vibration patterns. It can also create different sound sensations. Designers use isolation mounts, damping materials, and structural stiffness to manage NVH. They also tune software. Software can ramp torque smoothly. Smooth torque control can reduce harshness. It can also reduce wheel slip. It can reduce drivetrain shock.

Packaging decisions also affect thermal management. The platform must route cooling lines. It must place radiators, pumps, valves, and heat exchangers. It must also manage heating in cold climates. Batteries prefer a temperature window for best performance. Motors and inverters also have temperature limits. Cabin heating also consumes energy. Many EV designs use heat pumps or advanced HVAC strategies. These are platform-level choices because they affect space, cost, and manufacturing.

Another packaging topic is serviceability. A car must be maintained and repaired. Some designs make battery removal easier. Some make it harder. Some place the power electronics where they are accessible. Some place them in tight areas. Serviceability matters for warranty cost and customer experience. A design that saves a little space but creates expensive labor can be a bad business decision. Companies learn this over time through field data.

Platform design also interacts with product strategy. Some brands create dedicated EV platforms. Some adapt existing fuel platforms. Adapting can reduce early cost. It can also reduce time to market. But it can create compromises. Dedicated platforms can optimize packaging and efficiency. But they require larger investment. Neither approach is automatically better. The best approach depends on company goals, supply chain, and market timing.

A simple example helps. Imagine a small city EV designed for short trips. It can use a smaller battery. It can prioritize cost and easy parking. It can tolerate slower charging. Now imagine a long-distance highway EV. It needs a larger battery and better aerodynamics. It must manage heat during fast charging. It must handle high-speed stability. The platform choices will be different. The second car may need stronger thermal systems, and stronger high-speed aerodynamics. The first car may focus on low cost and simplicity.

Common myths appear in platform discussions. One myth is that a skateboard platform guarantees safety. Safety is a full system. It depends on structure design and validation. Another myth is that lower center of gravity always means a better car. It helps stability, but it is not the only factor. Another myth is that EVs have no vibration issues because they have no engine. EVs still have vibration sources. They are simply different sources.

Another common misunderstanding is about "empty space." People see an EV has fewer mechanical parts and assume there is unlimited space. In reality, the battery pack and thermal systems use significant space. High-voltage safety rules also require spacing and isolation. Wiring harnesses, controllers, sensors, and shielding also take space. Platform packaging is still a constrained puzzle.

The most important mindset for this chapter is systems thinking. Platform design is not a design sketch. It is a set of linked choices. Each choice affects cost, range, safety, NVH, charging, and service. Good platforms make later decisions easier. Poor platforms create constant compromise.

Quick Check

1. What does "platform" mean in EV design, in simple terms?

2. Why can a low battery pack improve stability, but still create trade-offs?

3. What is weight budgeting, and why is it critical for EVs?

4. Why can EVs reveal more noise compared to fuel cars?

5. What is the purpose of battery enclosure strength in a crash?

6. Why does serviceability matter for business outcomes?

7. What is one reason dedicated EV platforms can outperform adapted ones?

Chapter 13 — Battery Manufacturing and Supply Constraints

The battery is the largest cost driver in many EVs. It is also one of the hardest parts to manufacture consistently. A battery cell is not only a container of energy. It is a carefully built electrochemical system. It needs precise materials. It needs clean production. It needs tight control of moisture and contamination. Small defects can create big problems later. That is why battery manufacturing is both a chemistry process and a quality process.

When people talk about "cell production steps," they often imagine one simple factory line. In reality, cell production is a chain of steps. Each step has quality risks. Each step can create defects if control is weak. The exact steps vary by cell type and manufacturer. This chapter stays high level. It focuses on what is stable across most modern lithium-ion style production.

A cell starts with materials. The main materials include cathode material and anode material. They also include electrolyte, separator, and current collectors. The exact chemistry can differ. Some cathodes use different metals. Some anodes use different mixes. But the manufacturing discipline stays similar. Materials must meet purity and consistency standards. The factory must track batches. If batch quality changes, performance can change. This is one source of cost fluctuation. Materials markets can change, and quality can change.

Electrode production is a core stage. Electrode production generally involves coating active materials onto metal foils. The coating must be uniform. Thickness must be controlled. Then the coated electrodes are dried. Drying must remove solvents properly. If drying is poor, the electrode can have weak performance. It can also have safety risks. After drying, electrodes are pressed and calendered. Calendering means compressing to a target density. Density affects energy capacity and resistance. The electrodes are then cut into shapes.

Next comes cell assembly. Cells can be cylindrical, prismatic, or pouch. The packaging differs, but the assembly logic is similar. Electrodes and separator layers are stacked or wound. Then tabs are connected. The cell is sealed. At this point, moisture control is critical. Many processes happen in dry rooms. Moisture can cause chemical problems. It can also create gas and performance loss. It can also increase defect risk.

Then electrolyte is filled into the cell. Electrolyte filling must be precise. Too little or too much can cause issues. Then the cell often goes through formation. Formation is a controlled first charging process. It helps create the solid electrolyte interface layer, often called the SEI. This layer is part of normal lithium-ion behavior. It forms early and then stabilizes. Formation is slow and energy-intensive. It is also a quality filter. Cells that behave abnormally can be identified during formation. So formation is both a process and a test.

After formation, cells are aged and tested. They are measured for capacity and internal resistance. They are also checked for leakage and self-discharge. Self-discharge means the cell loses charge without use. High self-discharge can signal internal defects. Cells are then graded. Grading means sorting by performance. Higher grade cells may be used for higher performance packs. Lower grade cells may be used for less demanding applications. If yield is low, costs rise. If yield is high, costs fall.

Quality control checkpoints exist at every stage. Some checkpoints are simple measurements. Some are advanced inspections. The exact tools vary. But the mindset stays the same. Detect defects as early as possible. Early detection is cheaper. If defects are found late, you scrap more value. A defect that passes through many steps wastes materials and labor. It also reduces capacity and drives up unit cost.

Defect risks can be small and hard to detect. Particle contamination can cause internal short circuits. Poor coating uniformity can cause hotspots. Misalignment can cause local stress. Separator damage can cause shorts. Poor sealing can allow moisture entry. Each defect can be rare, but even rare defects matter at scale. When you ship millions of cells, small defect rates become real events.

This is where yield and scrap matter. Yield is the percent of product that meets quality standards. Scrap is what is rejected. Yield affects cost strongly. A factory with 95 percent yield is very different from one with 85 percent yield. The cost difference can be large. Yield also affects supply availability. If yield drops, you may not meet delivery plans. Then the vehicle factory can face production constraints. This is one reason EV supply can be unstable.

Costs fluctuate because materials prices can change. Costs also fluctuate because yield can change. Yield can change due to process tuning. It can change due to new equipment. It can change due to new chemistry. It can change due to supplier variability. When new cell lines are launched, yield can be low at first. Over time, yield often improves as processes stabilize. That improvement can reduce costs. But improvements can be slow and uncertain.

Supply constraints are not only about cells. They also include pack components. Packs need modules, busbars, sensors, cooling plates, and structural parts. They need electrical connectors and contactors. They also need battery management electronics. Any bottleneck can reduce pack output. A pack is a system of parts and processes. If one supplier is late, the whole line can slow.

Supply chain bottlenecks can be conceptualized as "single points of failure." If a critical material has few suppliers, risk rises. If a factory relies on a single process tool, downtime can halt production. If a shipping route is disrupted, deliveries can delay. EV programs therefore care about redundancy and flexibility. They also care about localizing supply when possible. Localization can reduce shipping risk. But it can also increase cost if scale is smaller.

Another supply topic is quality consistency across suppliers. Two suppliers can claim the same specification, but real performance can differ. A stable supply chain requires strong incoming inspection. It also requires strong supplier quality management. That includes audits and process capability checks. It also includes clear change control. Change control means a supplier cannot change a process or material without formal approval. Without change control, hidden changes can cause hidden defects.

One of the most important points is that batteries are made in controlled environments. Battery factories require dry rooms and clean handling. They require careful safety design. They also require strict worker training. They require automation for repeatability. Manual handling can introduce variability. Automation can reduce variability, but it also brings maintenance needs and calibration needs. When automation drifts, defects can rise. So automation must itself be managed.

The EV world also depends on learning cycles. If a pack issue appears in the field, companies investigate root causes. Root cause work is not about blame. It is about systems improvement. They look for patterns. They trace production batches. They compare test results. They identify which step might have caused the defect. Then they adjust controls. Over time, this learning improves reliability. But it requires discipline and data.

A practical example shows this chain. A manufacturer experiences a rise in warranty claims related to reduced range. They find that affected vehicles share a cell batch window. They investigate the electrode coating process. They find that a coating thickness drifted due to a maintenance issue. They then update maintenance schedules. They add a tighter measurement checkpoint. They also adjust acceptance thresholds. Over time, yield improves and claims fall. This is how quality systems create business value.

Common mistakes in public discussion include assuming battery supply is only about "mining." Mining is one part. Processing and production quality are also major parts. Another mistake is assuming every new cell chemistry will scale quickly. Scaling is hard. Another mistake is assuming cost declines are guaranteed. Costs can decline, but they depend on learning, yield, and materials markets. Another mistake is assuming defects are always obvious. Many defects are subtle and appear later.

A calm conclusion for this chapter is straightforward. Battery manufacturing is a precision industry. It rewards control and learning. It punishes shortcuts. When people understand this, they understand why battery supply can shape EV availability and price. They also understand why quality systems in battery plants are so important.

Quick Check

1. Why is battery manufacturing a "precision" process?
2. What is formation, and why is it both a process and a test?
3. Why do small defect rates matter so much at large scale?
4. What is yield, and how does it affect cost and supply?
5. Why can supply constraints come from pack parts, not only cells?
6. What is change control, and why do manufacturers demand it?
7. What is one reason costs can fluctuate even when demand is stable?

Chapter 14 — Assembly, Testing, and Reliability

Building an EV at scale is not only assembling parts. It is building repeatable outcomes. The same car must behave the same way for most customers. It must start reliably. It must charge reliably. It must brake and steer predictably. It must be safe in normal use and in rare situations. Reliability is not a single test. It is the result of design, production, supplier quality, and software discipline working together.

Assembly has a simple goal. Put parts together correctly, every time. But "correctly" has many layers. Torque must be correct. Seals must be correct. Electrical connectors must be seated correctly. Cooling lines must not leak. Sensors must be aligned. Software must be configured. If any step fails, the vehicle can still leave the factory, but problems may appear later. That is why assembly needs test checkpoints.

EV assembly also has unique elements. High-voltage systems must be safe. Workers need safety procedures. Tools and fixtures must be insulated when needed. Battery packs can be heavy and require careful handling. Thermal systems can be more complex than in some fuel cars because they manage the battery, the cabin, and power electronics. Wiring harnesses can be large because of sensors and controllers. Assembly lines must manage this complexity without creating variability.

End-of-line testing is one of the most important steps. End-of-line means the vehicle is assembled and is now tested before shipping. End-of-line tests check that key functions work. They check that there are no critical leaks. They check that the vehicle communicates correctly. They check that sensors report correctly. They check that safety systems have no faults. They also often check that the vehicle meets basic performance targets.

End-of-line tests can include electrical safety tests. A key concept is insulation resistance. High-voltage systems must be isolated from the vehicle body. If insulation is weak, there is a safety risk. Many EVs monitor insulation during operation. But end-of-line testing ensures the vehicle starts within safe thresholds. Another test is high-voltage interlock. Interlock systems detect if connectors are not properly closed. If interlock is broken, the system can prevent operation. This is part of safety-by-design.

Thermal tests can also occur. Cooling pumps can be tested. Valves can be actuated. Temperature sensors can be verified. In some factories, a short drive test is performed. The vehicle may be driven on rollers. That test checks acceleration, regen, brakes, steering, and noise. Some factories use automated test tracks. The goal is not to simulate years of driving. The goal is to catch assembly defects early.

Reliability also depends on software validation. Modern EVs are software-defined in many functions. Software controls battery charging. Software controls torque delivery. Software controls thermal behavior. Software also controls driver displays and user features. Software defects can create safety risks and customer dissatisfaction. So software must be validated. Validation means testing to ensure software does what it should do, under many conditions.

Software validation is difficult because real life has many edge cases. Cold weather can expose range estimation errors. Fast charging can expose thermal control limits. Weak cellular coverage can expose update system failures. Sensor noise can expose control loop instability. Good software teams test under varied conditions. They also collect field data. Field data is used to improve future versions. But field data must be handled carefully to respect privacy and security rules.

Update discipline is another key concept. Over-the-air updates can improve vehicles. They can fix bugs and improve performance. But updates also carry risk. A bad update can introduce new defects. It can disable features. It can create user confusion. That is why update discipline matters. A disciplined update process includes staged rollouts. It includes rollback plans. It includes clear release notes. It includes strong testing before release. It also includes careful version control.

Warranty realities provide another layer of learning. Warranty claims are not only costs. They are signals. They show where the system is weak. A mature organization treats warranty data as an improvement engine. They classify claims by type. They link claims to production batches when possible. They look for patterns by region and climate. They also look for patterns by usage type. Then they run root-cause analysis.

Root-cause learning means identifying why the problem happened, not only what the problem is. If a connector fails, the cause might be improper torque. It might be a tool calibration issue. It might be a supplier material issue. It might be a design tolerance issue. It might be vibration behavior. Root-cause work often shows that problems are multi-factor. Fixing one layer might reduce the issue, but not eliminate it. This is why system thinking matters.

This is where the Six Sigma mindset helps, when used practically. Six Sigma is often used as a buzzword. Here we keep it simple. The practical mindset is to reduce variation. Variation is the enemy of reliability. If the same process produces different outcomes, you get unpredictable vehicles. Reducing variation starts with defining critical-to-quality characteristics. These are the few outcomes that matter most. For EV assembly, examples include torque specs on critical joints, leak rates on cooling lines, insulation resistance, connector seating, sensor calibration, and software configuration integrity.

Then you measure these characteristics. You track them over time. You look for drift. Drift can occur due to tool wear, worker changes, or supplier variation. If you detect drift early, you fix it before it becomes a field problem. This is process control. It can be simple charts and thresholds. It does not need complex words to be effective.

Defect reduction also uses error-proofing. Error-proofing means designing the process so mistakes are hard to make. For example, fixtures can prevent wrong part orientation. Tools can confirm correct torque and record results. Connectors can be keyed so they cannot be inserted incorrectly. Software can validate configuration before shipping. These controls add cost, but they often reduce warranty costs more than they cost. They also protect brand trust.

Reliability is also about supplier management. Many EV parts come from suppliers. Batteries, motors, inverters, semiconductors, connectors, sensors, and software modules can come from different organizations. If suppliers change processes without notice, defects can increase. This is why supplier quality systems matter. It is also why manufacturers use audits and strict acceptance tests.

Another key reliability topic is "infant mortality." This term means early-life failures. Some defects appear quickly after production, often due to assembly defects or hidden component defects. Good end-of-line testing reduces infant mortality. But some issues still appear in the first months. Organizations often track early failures separately. They treat them as urgent signals that a process may be out of control.

There is also "wear-out" reliability. Some failures appear after long use. These can relate to thermal cycles, vibration, corrosion, and repeated charging. EVs have different wear patterns compared to fuel cars. For example, regenerative braking can reduce brake pad wear, but it can also cause less brake use, which can lead to corrosion in some climates. Battery performance can degrade over time. Cooling systems can age. Power electronics can face thermal stress. Predicting these patterns requires long-term testing and field data.

A practical example can connect these ideas. Imagine a manufacturer sees a rise in charging complaints. Some customers report slower charging than expected. The company checks vehicle logs and sees that thermal limits are triggering early. They then look at assembly records and find that a coolant line clamp torque varies. A small leak leads to low coolant performance. That reduces thermal capacity. That reduces charging performance. They fix the clamp process. They add a leak test threshold. Complaints drop. This chain shows how a small assembly issue can create a customer experience issue that looks like a "technology problem."

Common mistakes in public discussion include assuming end-of-line testing can "guarantee" long-term reliability. It cannot. It can reduce early failures. Another mistake is assuming software updates are always good. They can be good, but they require discipline. Another mistake is assuming warranty problems mean the whole technology is flawed. Sometimes they reflect a specific process issue that can be fixed. Another mistake is assuming EVs have fewer parts, so quality is easy. EVs may have fewer mechanical parts in some areas, but they have more electronics and software complexity. Complexity shifts rather than disappears.

A calm conclusion for this chapter is clear. High reliability comes from discipline. It comes from stable processes, good testing, careful updates, and honest learning from failures. It is not glamorous. It is what makes an EV feel trustworthy in daily life.

Quick Check

1. What is end-of-line testing, and what is its main purpose?
2. Why are insulation and interlock checks important in EVs?

3. Why does software validation matter for safety and reliability?

4. What does update discipline mean in practice?

5. Why are warranty claims useful signals, not only costs?

6. What does "reduce variation" mean in simple quality terms?

7. How can a small assembly issue create a big user experience problem?

Part 5 — Markets, Business Models, and Competition

EVs are not only machines. EVs are a market system. Value is created in many places. Value is also captured in many places. Some companies earn from hardware. Some earn from software. Some earn from charging services. Some earn from energy retail. Some earn from financing and warranties. Some earn from fleets and service contracts. This part explains where money is made, and why. It also explains why adoption moves in waves. It explains why prices can stay high even when technology improves. It explains why trust becomes a competitive advantage. The goal is not to predict winners. The goal is to build a clear mental map you can use. You will learn how to read the EV industry without noise.

Chapter 15 — The EV Value Chain (Who Makes Money Where)

An EV reaches a driver through a long chain. That chain starts with raw materials. It ends with an operating vehicle on the road. Every step has costs. Every step has risks. Every step can also create profit. When you understand the chain, you understand competition. You also understand why some EV companies grow fast, then slow. You also understand why some suppliers become powerful. You also understand why charging matters economically.

The value chain begins with automakers, but it does not belong to them alone. Automakers assemble the final product. They design the platform. They integrate parts. They manage safety validation. They build a brand promise. They also manage distribution and service networks. Their profits come from selling vehicles. They also come from financing in many cases. They also come from after-sales services. After-sales includes maintenance, parts, and repairs. In EVs, after-sales can be smaller in some areas. But it can be larger in others. Software and electronics can create new service needs.

Battery makers are central to the chain. Batteries are high-cost components. They also influence performance. They influence range and charging. They influence safety and durability. Battery makers earn from cell sales and pack sales. Their profitability depends on yield and scale. It also depends on material prices. It depends on quality control. It depends on long-term supply contracts. Some automakers make batteries themselves. Some partner with specialists. Some split designs and outsource production. There is no single model that is always best. The best model depends on strategy and capability.

Suppliers are a large part of the EV economy. Suppliers provide motors, inverters, power modules, semiconductors, sensors, wiring, thermal systems, and connectors. Many of these parts are specialized. Some have high barriers to entry. That can give suppliers pricing power. It can also give them negotiation strength. At the same time, suppliers face intense quality demands. A small defect can lead to large recalls. So suppliers invest heavily in quality systems. They also invest in process stability.

Semiconductors deserve a clear mention. EVs can use more power electronics and control electronics than some older vehicle designs. That can increase dependency on chip supply. Chips also have lead times. Shortages can slow production. When chips are tight, the chain becomes fragile. Companies respond by redesigning control units. They also sign long-term contracts. They also diversify suppliers. This is a strategic topic, not only an engineering topic.

Charging operators and energy retailers form another layer. Charging is both infrastructure and a service business. Charging operators may own and operate charging stations. They pay for site development and equipment. They pay for maintenance and customer support. Their revenue comes from charging sessions. Their costs include electricity, rent, repairs, and network operations. Profit depends on utilization. Utilization means how often chargers are used. If chargers sit idle, fixed costs remain. If utilization rises, the business improves. This is why location matters. It is why reliability matters. It is why trust matters.

Energy retailers and utilities also participate. In many places, electricity pricing and access are regulated in specific ways. The details vary by country and region. So we keep it high level. The stable idea is this. Electricity has wholesale costs and delivery costs. It also has peak demand costs. Charging can increase peak demand. That can raise costs for operators. Smart load management can reduce those costs. So software and operational discipline can directly influence profit. In other words, good operations can act like a "cost advantage."

Software platforms and services are growing in importance. Software can include vehicle operating systems, energy management systems, navigation and route planning, charging payment and roaming, fleet management dashboards, and predictive maintenance tools. Some software is inside the car. Some software is in the cloud. Some software connects both. Companies can earn subscription revenue. They can earn service revenue. They can earn from enterprise contracts. They can also earn from data services, but this area requires strict privacy discipline. Data value is real, but trust risk is also real.

Fleet services create another profit pool. Fleets care about uptime and predictable cost. Fleets often prefer service contracts. They may buy vehicles with maintenance bundles. They may also buy charging management as a service. They may also buy training and safety compliance support. In this model, value is captured through long-term relationships. It is less about one-time sales. It is more about lifetime customer value. This is why many EV companies focus on fleets early. Fleets provide concentrated demand. Fleets also provide structured feedback.

Recycling and second-life markets are a newer layer. The battery does not become useless when it is no longer ideal for driving. It may still hold useful capacity. It may be used in less demanding storage roles. This is sometimes called second-life use. The economics depend on testing, safety, and logistics. It also depends on regulation. It also depends on warranty and liability rules. These details vary widely, so we stay conceptual. The stable insight is this. Second-life and recycling can reduce waste. They can also recover materials. They can reduce supply risk in the long run. But they also require industrial systems that are still developing in many markets.

Recycling itself is not one single method. There are different industrial approaches. Outcomes depend on technology, scale, and local rules. The key business point is that recycling becomes more important as EV volumes grow. It can become a source of materials. It can also become a compliance requirement. It can also become a brand and trust factor. If customers believe a company handles end-of-life responsibly, that can support brand trust.

Now, who "makes money where" depends on who controls scarcity and complexity. If batteries are scarce, battery makers gain power. If software creates strong user lock-in, software platforms gain power. If charging is scarce and reliable networks are rare, charging operators can gain power. If manufacturing scale and quality are hard to replicate, automakers with strong operations can gain power. The chain is dynamic. It changes as technology and infrastructure mature.

A useful way to see the chain is to ask simple questions. Who owns the customer relationship. Who owns the charging experience. Who owns the data and user identity. Who owns the critical components. Who has the best cost control. Who has the strongest trust reputation. These questions reveal why competition is not only about range or acceleration.

Consider a practical example. A city delivery fleet wants to switch to EVs. They need vehicles, charging, and maintenance. They may buy from an automaker, but they also need chargers. They may outsource charging operations to a service provider. They may use a software platform to manage routes and charging schedules. Each participant captures value. The automaker earns from vehicle sale and service contract. The charging operator earns from energy and uptime service. The software provider earns from subscription. The fleet chooses the combination that reduces risk and cost.

A second example is a consumer buyer. The consumer sees a vehicle price. But they also face charging access. They also face insurance and resale value. They also face software support over time. If they trust the brand and the charging ecosystem, they buy. If they do not trust it, they delay. The value chain is therefore linked to trust at the end.

Common misconceptions create confusion. One misconception is that automakers must capture all value. They do not. Another misconception is that charging is "just electricity." It is not. It is also maintenance, software, and site operations. Another misconception is that software value is always easy. It is not. Software requires continuous updates and security work. Another misconception is that recycling will quickly become a major profit pool. It may grow, but its pace depends on scale and policy and safety systems.

The calm conclusion is simple. The EV industry is a network of business models. It is not one business. Understanding this helps you read headlines correctly. You can see where value is real. You can also see where claims are only marketing.

Quick Check

1. Why do batteries have such influence on EV economics?

2. What makes charging a service business, not only a hardware business?

3. How can software create recurring revenue in EV ecosystems?

4. Why do fleets often adopt EVs differently than consumers?

5. What does utilization mean for charging network profitability?

6. Why is the supply chain a strategic risk in EV production?

7. Name one reason recycling and second-life are "system problems."

Chapter 16 — Pricing, Total Cost of Ownership, and Adoption Drivers

EV pricing often feels confusing. Many people ask a simple question. Why is the purchase price still high in many cases. The answer is not one thing. It is a set of factors. Battery cost is one factor. Manufacturing scale is one factor. Supply chain is one factor. Software and electronics content is one factor. Brand strategy is also a factor. Policy and incentives can also shape price signals. But incentives vary by location and time, so we treat them as high level concepts.

Total cost of ownership, often called TCO, is a more complete lens than purchase price alone. TCO includes what you pay up front. It also includes energy cost over time. It includes maintenance and repairs. It includes insurance and taxes in many markets. It includes resale value. It also includes your time and convenience costs. Convenience costs can include charging access and downtime. Time is real value, even if people do not label it as "cost."

Purchase price is the most visible number. For many buyers, it is the barrier. For fleets, purchase price is only one input. They often look at cost per mile or cost per kilometer. They look at downtime. They look at maintenance schedules. They look at driver productivity. They also look at financing terms. A fleet may accept a higher purchase price if operating costs are lower and uptime is high. A consumer may not accept that if the monthly payment is too high.

Lifetime cost depends heavily on energy cost. If electricity is cheap and stable, EV operating cost can be attractive. If electricity is expensive or unpredictable, savings can shrink. It also depends on charging behavior. Home charging tends to be cheaper and easier in many settings. Public fast charging can be more expensive. It can also be less convenient if stations are busy or unreliable. So the same EV can have different operating costs for different people. This is why "EVs are always cheaper" is not always true. It depends.

Maintenance differences are often oversimplified. EVs can have fewer routine mechanical maintenance items. They do not need oil changes. They may have less brake wear due to regenerative braking. They also have fewer parts in some powertrain areas. But EVs still need tires, suspension work, alignment, filters, coolant service in some designs, and brake fluid service. They also have software issues and electronics issues. They also have thermal system complexity. So the correct statement is balanced. Some maintenance categories can shrink. Other categories remain. Some new categories appear.

Misconceptions around maintenance can harm adoption. Some people assume EVs need almost no service. Then they are surprised by tire wear or by calibration needs. Other people assume EVs are fragile and expensive to fix. That can also be wrong. The truth is mixed and depends on design and service network quality. Service network quality matters because repair time and parts availability affect downtime. Downtime is a cost, especially for fleets.

Fleet economics differ from personal economics in several ways. Fleets can centralize charging. They can negotiate electricity contracts. They can schedule charging off-peak. They can install load management. They can spread fixed costs across many vehicles. They can also train drivers. Training reduces damage and improves efficiency. Fleets can also track data and adjust operations. A consumer may not have these advantages. A consumer may rely on public charging. A consumer may have fewer tools to plan energy costs. So fleets can sometimes reach positive TCO earlier than consumers.

Personal economics involve lifestyle fit. If you can charge at home, EV ownership can feel easy. If you cannot charge at home, EV ownership can still work, but it requires infrastructure and routine. That routine can add stress. The "cost" of stress is real. People avoid systems that feel risky. This is why public charging reliability can influence adoption as much as sticker price. A cheap EV with poor charging experience can still feel like a bad deal.

Another adoption driver is resale value. Resale value is influenced by brand trust, warranty terms, battery health perception, and software support. Battery health perception matters because buyers worry about range loss. Some sellers can provide battery health reports. That can reduce fear. But standards differ and availability varies. So we keep it conceptual. The stable point is this. Transparent battery health information can support resale trust. Unclear information can reduce resale prices. Lower resale prices can increase total cost of ownership.

Warranty and reliability also matter. If buyers believe repairs will be slow or expensive, they demand a lower price. If buyers trust the service network, they pay more. Trust therefore becomes a pricing factor. It becomes a "hidden asset" that allows stronger margins. This is true in many industries, not only EVs.

Incentives and policy effects can influence adoption. Some regions offer purchase incentives or tax benefits. Some offer non-monetary benefits, like access privileges. Some also apply emissions rules that influence automaker strategy. But these policy choices vary and can change. So the stable guidance is simple. Treat incentives as temporary support, not as permanent business fundamentals. A strong EV product should still make sense without assuming long-term subsidies.

Policy can also shape charging infrastructure. Public investment or regulation can speed up deployment. Interoperability rules can reduce payment friction. Building codes can enable chargers in new buildings. These actions influence adoption by reducing inconvenience. Again, details vary. The stable principle is that infrastructure and standards can lower adoption friction.

A practical way to evaluate TCO is to build a simple scenario. You estimate annual miles. You estimate average energy use. You estimate electricity price for your main charging source. You estimate expected maintenance and tires. You estimate insurance and fees. Then you compare to an alternative vehicle. But you must also include charging access and downtime risk. For example, if you depend on public fast charging and it is unreliable, you may waste time. That time has value. A fleet manager counts that time. A consumer feels that time as frustration.

Let's use a calm example. A commuter drives mostly in a city and can charge at home. They use overnight electricity. Their routine is stable. Their driving is predictable. For this person, EV adoption can be driven by convenience and predictable fuel cost. Now consider a person who lives in an apartment with no home charging and drives long trips often. For this person, the decision depends more on public charging reliability and travel patterns. The same EV can be a great product for one person and a stressful product for the other.

Another example is a delivery business. They have fixed routes and return to base. They can plan depot charging. They can manage peaks. They can minimize downtime. For them, EV adoption can be driven by predictable operating cost and reduced mechanical downtime. But they must plan charging capacity and maintenance training. If they skip planning, failures happen. Then the EV looks "bad," but the root cause is system design, not the vehicle alone.

Common mistakes include focusing only on purchase price and ignoring the system. Another mistake is comparing EV range numbers without considering charging access. Another mistake is assuming all electricity is cheap. Another mistake is assuming all public charging is expensive. Another mistake is treating incentives as guaranteed. Another mistake is ignoring resale. Another mistake is ignoring insurance differences, which can vary widely.

A calm conclusion is this. EV economics are personal and operational. They depend on your charging access, your driving pattern, and your risk tolerance. TCO thinking helps you avoid being misled by a single number. It also helps you compare products without emotion.

Quick Check

1. Why is purchase price not the same as total cost of ownership?

2. What is one hidden cost that can matter for EV owners?

3. Why can fleets reach positive TCO sooner than consumers?

4. Why does home charging change economics and convenience?

5. How can resale value influence TCO even before you sell?

6. Why should incentives be treated as "context," not a guarantee?

7. What is one common misconception about EV maintenance?

Chapter 17 — Competition, Differentiation, and Brand Trust

EV competition is intense because the industry is still forming. Some companies compete on price. Some compete on performance. Some compete on software features. Some compete on charging ecosystems. Some compete on reliability and service. Many compete on all of these at once. But not every company can win in every dimension. So differentiation matters. Differentiation means how a product stands out in a way customers value.

A key issue in EV competition is the gap between claims and real usability. Range claims are a good example. Many buyers focus on the largest range number. But real usability includes more than range. It includes charging speed behavior across battery levels. It includes charging network access. It includes cold-weather behavior. It includes highway efficiency. It includes reliability. It also includes how confident you feel during daily use. A vehicle can have strong claimed range but still feel stressful if charging is inconsistent.

This is why "range claims vs real usability" is a useful lens. Real usability answers simple questions. Can you do your common trips without planning anxiety. Can you charge easily where you live and work. Can you travel long distance with predictable stops. Can you trust the car in heat and cold. Can you get service when needed. These are not marketing questions. They are life questions. Competitive advantage comes from delivering calm answers to these questions.

Charging experience is becoming a competitive factor because it shapes daily trust. Charging experience includes more than charger count. It includes reliability, simple payment, and accurate station status. It includes station placement and safety. It includes how fast the vehicle charges at typical battery levels. It includes how the car plans routes. It includes how clearly it communicates charging time. A good vehicle plus a weak charging experience can still lose. A weaker vehicle plus a strong charging experience can win for many buyers.

Competition also happens through quality and safety perception. Safety perception includes crash safety, but it also includes fire fear and software fear. Many people have strong emotions around batteries. Brands must respond with calm facts and clear guidance. They must also design safety systems and communicate them clearly. They must handle incidents with transparency. In a trust-based market, how a company responds to problems can matter as much as the problem itself.

Quality perception is built through consistency. Buyers do not want surprises. They want consistent panel fit, stable software, and predictable service. They also want the interior to age well. They want fewer rattles. They want fewer charging glitches. They want fewer warning lights. Quality is not only "luxury." Quality is reduced friction. Reduced friction makes a product feel premium even if it is not expensive.

Reputation is also influenced by service networks. A strong product can be damaged by weak service. Long repair wait times destroy trust. Parts shortages destroy trust. Poor communication destroys trust. In competitive markets, trust can be lost quickly. It can take years to rebuild. This is why serious EV companies invest in service readiness. They also train technicians. They also manage parts logistics. They also improve diagnostics tools.

Software also shapes differentiation. Some companies add many features. Features can be helpful. But features can also create bugs. So differentiation through software requires discipline. Stable software can become a brand advantage. Unstable software can become a brand risk. A company that updates carefully can improve customer experience over time. A company that updates carelessly can harm it. This affects loyalty and resale value.

Brand trust is not a soft idea. It changes business outcomes. It reduces customer acquisition cost. It increases repeat purchases. It improves word-of-mouth marketing. It improves fleet contract wins. It supports better resale values. It also supports pricing power. Pricing power means the ability to charge higher prices without losing demand. Trust is one of the strongest sources of pricing power in markets with risk and uncertainty.

Competition also involves cost control. Cost control is often invisible to customers. But it decides which companies can survive price wars. It decides who can offer lower prices without losing money. Cost control includes manufacturing efficiency, yield, supply chain contracts, and platform reuse. It also includes designing for manufacturability. A company can make a technically excellent car and still fail if it cannot build it efficiently at scale.

Another competitive axis is ecosystem lock-in. Ecosystem lock-in means a user stays because the whole system works together. This can include app ecosystems, charging memberships, and fleet management tools. Lock-in can help companies keep customers. But it can also create backlash if it feels unfair. The best ecosystems provide value without trapping users. They compete through convenience rather than restriction.

Now, how should a buyer compare products without marketing noise. The first step is to define the job-to-be-done. A job-to-be-done is the real purpose of the vehicle for that buyer. A commuter job is different from a road-trip job. A family job is different from a delivery job. A city job is different from a rural job. When you define the job, you can filter marketing.

Then you focus on a few stable criteria. For example, charging access, real driving range in your pattern, service coverage, safety reputation, and software stability. You can also include build quality and warranty clarity. These criteria matter across brands. They also matter across price tiers. The correct comparison is not "best EV." It is "best EV for this job and this environment."

A practical example can illustrate this approach. Two vehicles have similar range claims. Vehicle A has better highway efficiency and stable route planning. Vehicle B has faster peak charging but weaker charging network access. A highway traveler may prefer Vehicle A because predictability matters more than peak speed. A local commuter with home charging may prefer Vehicle B because peak charging rarely matters. The "winner" depends on context.

Another example is fleet procurement. A fleet may prioritize uptime, service response time, and total cost. They may accept lower performance because performance does not create business value. They may prioritize data tools and driver training support. A consumer buyer may value comfort and design more. Again, the competitive landscape is segmented by jobs and buyers.

Common mistakes include chasing the largest numbers and ignoring the system. Another mistake is trusting marketing claims without checking the operational reality. Another mistake is assuming one brand's strength applies to every use case. Another mistake is ignoring service and parts availability. Another mistake is ignoring charging experience. Another mistake is making decisions based on social media hype.

A calm conclusion is this. EV competition will not be decided by one metric. It will be decided by trust and system quality. The most successful companies will likely be those that reduce friction across the full ownership journey. That journey includes buying, charging, driving, updating, servicing, and reselling.

Quick Check

1. What is the difference between range claims and real usability?

2. Why is charging experience a competitive factor beyond charger count?

3. How can service network quality influence brand reputation?

4. Why can software features become a risk if update discipline is weak?

5. What does pricing power mean, and why does trust support it?

6. Why is "best EV" the wrong question without a use case?

7. Name one common mistake people make when comparing EVs.

Part 6 — Regulation, Safety, and Responsible Adoption

Electric vehicles do not grow only because the technology improves. They grow because the system around them becomes safe, trusted, and workable. Rules and standards help create that system. They reduce uncertainty for buyers, companies, and governments. They also reduce risk for the public. When rules are unclear, investment slows. When rules are consistent, adoption becomes easier.

This part stays high level on purpose. Laws and standards differ by country and region. They also change over time. The goal here is not legal guidance. The goal is understanding. You will learn why regulators care, what safety testing tries to prove, and why charging standards matter. You will also learn why data responsibility is now part of "vehicle safety." Finally, you will learn how public trust is built, and how it is lost.

Chapter 18 — Regulation and Standards (Global View, High Level)

Rules exist because mobility affects everyone. A vehicle shares roads with other vehicles. It shares spaces with people. It uses public infrastructure. It can create harm if it fails. That harm can be physical, financial, or digital. Regulation is the tool societies use to reduce that harm. Standards are the tool industries use to make products consistent and safer. These two ideas work together, but they are not the same.

Regulators care because EV adoption changes risk patterns. Batteries store large amounts of energy. High voltage systems require new safety thinking. Software updates can change vehicle behavior after sale. Charging connects vehicles to the electrical grid and to public spaces. That creates new points of failure. Regulators want to reduce accidents, reduce fraud, and protect the public. They also want fair markets. They want clear responsibility when things go wrong.

It helps to separate three layers of regulation. The first layer is vehicle approval and roadworthiness. This includes basic safety requirements. It includes crash safety concepts and electrical safety concepts. It also includes labeling and consumer information rules. The second layer is charging and infrastructure. This includes installation rules, site safety, and accessibility requirements. It can include payment transparency and reliability expectations. The third layer is data and software responsibility. This includes privacy expectations, security expectations, and incident reporting expectations in some markets.

Many people think regulation blocks innovation. Sometimes it can slow things down. But regulation can also enable growth. When consumers trust that a vehicle meets a safety baseline, they are more willing to adopt. When investors trust the rules, they invest more. When manufacturers know what to build toward, they can scale faster. The best regulatory systems balance safety with practical innovation.

Standards are different from laws. Standards are shared technical agreements. They define tests, methods, and interfaces. They help companies build compatible products. They also help buyers compare products. Standards can be voluntary or tied to regulation. In many places, regulation references standards. This means standards become the practical way to prove compliance. Standards also reduce confusion in supply chains. A supplier can design to a known standard. An automaker can integrate parts with fewer surprises.

Safety standards and testing concepts can be understood as proof of control. Testing asks a simple question. Under expected conditions, does the system remain safe. Under abnormal conditions, does the system fail in a predictable, limited way. EV safety is not only about avoiding failure. It is also about limiting damage when failure happens. This is sometimes called a "fail-safe mindset." It means the system tries to protect people even in bad scenarios.

Battery safety testing often focuses on heat, impact, and electrical faults. Heat matters because batteries are sensitive to temperature. Impact matters because crashes can damage packs and wiring. Electrical faults matter because high voltage must be isolated and controlled. Tests try to ensure that insulation remains effective. They try to ensure that the vehicle can shut down power safely. They also try to ensure that warning systems work when something is wrong.

Another safety concept is functional safety. This is a way of thinking about electronic and software systems. It asks how systems behave when sensors fail, when signals are wrong, or when software behaves unexpectedly. The goal is to reduce the chance that a failure leads to harm. In EVs, functional safety applies to power control, braking coordination, steering support systems, and thermal control. It also applies to battery management. A battery management system estimates state of charge and health. It also monitors temperature and voltage. Errors here can lead to poor decisions. So the system must be designed to detect faults early and respond safely.

Charging standards and interoperability matter because charging is a shared ecosystem. A driver expects a charger to work. A charger expects a vehicle to communicate correctly. Interoperability means different vehicles can charge at different stations without surprises. When interoperability is weak, adoption slows. People fear getting stuck. Fleets face downtime risk. Businesses hesitate to invest.

Interoperability is not only a plug shape. It also includes communication between the vehicle and charger. It includes how charging starts and stops. It includes how safety checks are performed. It includes how billing and access are handled. It also includes roaming, which is the ability to use one account across networks. These topics sound like "convenience," but they also affect safety. For example, unclear communication can cause charging to fail. It can cause overheating if systems do not coordinate properly. So standards aim to reduce both friction and risk.

A common misunderstanding is to treat standards as a guarantee of perfect experience. Standards reduce problems, but they do not remove them. Real-world reliability also depends on maintenance, software quality, and site conditions. A charger can meet a standard and still fail if it is not maintained. A vehicle can meet a standard and still confuse users if the interface is unclear. Standards are necessary, but they are not sufficient.

Data, privacy, and software responsibility have become part of EV regulation because EVs are connected products. They collect and process data to function. They also update over time. This creates two responsibilities. The first is privacy responsibility. Vehicles can record location data and usage patterns. That data can be sensitive. A responsible system limits data collection to what is needed. It also protects that data. It also gives users clear choices when possible. Rules vary, but the principle is stable.

The second responsibility is cybersecurity responsibility. Connected vehicles face cyber risks. Risks include account takeovers, fake apps, and unauthorized access. Risks also include spoofing and tampering with signals in some contexts. Not every risk is likely in every market. Some risks are high level and depend on threat actors. But the calm principle remains. Security must be designed, not added later. Security is also operational. It requires updates, monitoring, and responsible disclosure processes.

Software updates bring another responsibility. Updates can improve safety. They can fix defects. They can also introduce new defects if done carelessly. So responsible update discipline matters. A safe approach includes staged rollouts, testing, and clear release notes. It also includes a way to roll back if needed. It includes monitoring for unexpected behavior after release. Consumers do not need to know the technical details. But they deserve a product that does not change unpredictably.

Regulation also shapes business behavior. When compliance costs are high, small companies may struggle. When rules are unclear, companies delay launches. When rules are strict but predictable, companies can plan. This is why "global view" matters. A company may build for one region and adapt for another. That adds cost. It also increases complexity. It can slow down product cycles.

For buyers, the practical question is simpler. How do these systems protect me. The answer is that regulation and standards create baselines. They do not remove all risk. But they reduce common failure modes. They also create accountability. Accountability means someone must respond when systems fail. It also means companies must document and test. This documentation is not exciting. But it is part of what makes EV adoption sustainable.

One useful way to stay realistic is to avoid absolute statements. No product is risk-free. No regulation covers every situation. No test can prove every future scenario. But strong systems reduce risk to a level society accepts. That is the purpose. It is also why public trust can rise over time.

Common mistakes happen when people treat regulation as a simple "yes or no." Reality is more nuanced. A product may be legal but still inconvenient. A product may meet a standard but still have reliability problems. A product may be safe in design but unsafe in use if people ignore basic habits. This is why regulation and standards must be combined with education, training, and operational discipline.

Quick Check

1. Why do regulators care about EVs beyond normal road safety?

2. What is the difference between regulation and standards?

3. Why does interoperability matter for charging adoption?

4. How can software updates improve safety and also create risk?

5. Why are privacy and cybersecurity now part of "vehicle responsibility"?

6. Why can standards reduce problems but not guarantee perfect reliability?

7. What does accountability mean in the context of EV systems?

Chapter 19 — Safety, Risk, and Public Trust

Public trust is the real limit on EV adoption. Technology can improve quickly. Factories can scale. Charging can expand. But if people do not trust safety, adoption slows. Trust is shaped by experience, stories, and incident response. It is also shaped by how calmly risks are communicated. The goal is not to hide risk. The goal is to frame risk correctly, and reduce it through good practice.

Battery fire realities are often misunderstood. Batteries can burn under certain failure conditions. That is a fact. But the existence of a risk does not tell us its size or its context. Many daily systems have risks. Gasoline is flammable. Homes have electrical fires. Phones and laptops can overheat. The honest approach is to treat EV battery fires as a safety engineering problem. It is not a marketing battle. It is also not a reason for panic.

A calm risk framing starts with how fires occur. A battery fire is usually the end of a chain of failure. That chain may include severe physical damage, manufacturing defects, poor thermal control, or electrical faults. It may also include poor repair practices after a crash. It may include misuse of charging equipment. It may include exposure to extreme conditions combined with faults. Not every incident fits one story. Some causes remain unknown publicly. So we should avoid overconfident explanations when details are not confirmed.

The second part of calm framing is how risk is managed. EVs use multiple layers of protection. They use thermal management systems. They use battery management systems. They use fuses and contactors to disconnect power. They use insulation and isolation monitoring. They use structural protection around the pack. They use warning systems and limp modes. These layers do not make failure impossible. They aim to make failure less likely, and less harmful.

The third part is what owners and operators can do. Safety is partly design, and partly behavior. Owners can reduce risk by using approved charging equipment. They can avoid damaged cables. They can avoid unsafe adapters. They can keep charging areas clean and ventilated. They can follow storage guidance when a vehicle is unused for long periods. Exact guidance differs by model and region. So the stable principle is to follow manufacturer guidance and use reputable equipment. If guidance is unclear, treat it as a risk signal.

Workplace and fleet safety planning deserves focused attention. Fleets have repeated cycles of charging. They have many vehicles in one site. They have multiple drivers with different skill levels. This increases the need for structured discipline. A fleet safety plan should define roles, training, inspection routines, and incident response steps. It should also define who can handle charging equipment issues. It should define how damaged vehicles are isolated after collisions. It should define how battery warnings are escalated. The plan should be simple enough to follow, even under stress.

A strong fleet plan also includes site safety. Charging sites must consider spacing, ventilation, clear access routes, and basic fire readiness. It must also consider electrical load management. Overloaded circuits and improvised wiring are common sources of risk in many settings. This is not unique to EVs. It is basic electrical safety. In EV adoption, the scale makes it more important. A fleet that charges many vehicles at once must treat the site like an energy system, not like a normal parking lot.

Training is a major trust lever. Many incidents in transportation are linked to human factors. Human factors include distraction, fatigue, poor judgment, and rushed procedures. EVs introduce new procedures. Drivers must learn charging behavior and safe cable handling. Technicians must learn high voltage safety. Site managers must learn basic load and hazard awareness. If training is weak, small mistakes multiply. If training is strong, adoption becomes smoother and safer.

First-responder considerations matter because emergency teams need predictable guidance. After a crash, responders must secure the scene. They must manage fire risk. They must know how to disable systems safely. In many markets, manufacturers provide response guides. These guides explain safe cut points, isolation behavior, and towing considerations. The details vary. So the stable principle is that responders need clear information. They also need consistent labeling and access to guidance. The faster responders can act confidently, the safer the system becomes for everyone.

Public trust is also influenced by how companies and institutions respond to incidents. A poor response can damage trust even if the incident is rare. A strong response can protect trust even when an incident is serious. A strong response is transparent. It avoids blame. It explains what is known and unknown. It explains what actions are being taken. It also supports customers with clear steps. This is a trust discipline. It is not a public relations trick.

Common incident patterns can be described without fear. One pattern is charging misuse and poor equipment quality. Another pattern is damage after accidents that is not handled correctly. Another pattern is poor repairs by unqualified service providers. Another pattern is ignoring warning signs, like persistent error messages or unusual smells. Another pattern is unsafe storage practices in extreme heat or in enclosed spaces with poor ventilation. These patterns are not unique to EVs. But EV systems require attention because of energy density and high voltage.

Prevention habits should be practical and calm. Drivers should avoid charging in unsafe conditions. They should stop charging if equipment overheats or behaves oddly. They should not use damaged connectors. They should avoid water exposure beyond what equipment allows. They should keep charging areas clean. They should report repeated charging failures. They should not ignore warning indicators. These are basic habits. They reduce risk for individuals and for the public.

Public trust also depends on honesty about limits. EVs are improving, but they still face constraints. Some constraints are performance-related, like cold-weather range loss. Some constraints are infrastructure-related, like charger reliability. Some constraints are economic, like repair costs in certain markets. When companies hide these constraints, trust falls. When they explain them clearly, trust rises. Trust grows when expectations match reality.

Another trust factor is fairness and accessibility. If EV adoption feels like it serves only wealthy buyers, public support can weaken. If charging infrastructure is concentrated in rich areas only, public trust can weaken. If repair networks are limited, trust can weaken. Responsible adoption means building systems that work for wider groups over time. This includes fleets, shared mobility, public transport, and lower-cost segments as markets mature. This is not a moral lecture. It is a stability requirement for long-term industry growth.

Cybersecurity also affects public trust. People increasingly see vehicles as computers. If people fear hacking, they hesitate. The best response is not to promise "perfect security." Perfect security is not realistic. The best response is layered security and honest communication. Companies should protect accounts, secure updates, and respond quickly to discovered issues. Users should use strong passwords, avoid unofficial apps, and update software when advised. This shared responsibility model is similar to smartphones and online banking.

Responsible adoption also includes truthful marketing. Overstated autonomy claims, overstated range claims, and unclear warranty messaging can all damage trust. Trust is hard to gain and easy to lose. This is why the most sustainable companies often focus on reliability and clarity more than headlines. Headlines can create attention. Reliability creates retention. Retention creates stable growth.

In the end, safety is not a single feature. It is a system behavior over time. It includes design, manufacturing quality, software discipline, infrastructure reliability, training, and incident response. Public trust is the result of that system working well, repeatedly. This is why responsible adoption is not only a regulatory topic. It is a leadership topic. It is also an operational discipline topic.

Quick Check

1. Why is public trust a limiting factor for EV adoption?
2. What does "calm risk framing" mean in safety communication?
3. Why do fleets need structured safety plans more than individuals?
4. Name two common incident patterns that can be reduced by good habits.
5. Why do first responders need clear EV information after crashes?
6. How can cybersecurity affect safety perception and adoption?
7. Why does honest marketing support long-term trust better than hype?

Part 7 — Sustainability, Materials, and Lifecycle

Electric vehicles change what we see. They remove tailpipe smoke and noise. But they also move impacts upstream. Those impacts sit in mines, factories, power grids, and recycling systems. This part gives you a calm way to think about the full lifecycle. It does not use slogans. It does not pretend there is one perfect answer. It helps you ask better questions. It also helps you separate what is measurable from what is still uncertain.

Sustainability is not one score. It is a system trade-off. It depends on materials, energy, and time. It depends on where the vehicle is built. It depends on how it is used. It depends on how long it lasts. It depends on what happens when it retires. If you understand the chain, you can judge claims with discipline. You can also avoid common traps, like comparing two cars using only one number.

Chapter 20 — Materials, Mining, and Supply Ethics (High Level)

An EV is a product of many materials. Some materials are common. Some are specialized. Batteries and power electronics rely on supply chains that can be concentrated. This creates two realities at the same time. The first reality is that EVs can reduce local air pollution in cities. The second reality is that material extraction and processing can create harm elsewhere. Responsible thinking must hold both realities without denial.

Key materials matter because they shape performance and cost. A battery needs active materials that store and release energy. It also needs current collectors and separators. It needs a stable casing. It needs thermal interface materials. It needs control electronics and sensors. Beyond the battery, the vehicle uses copper in wiring and motors. It uses magnets in some motor designs. It uses aluminum and steel in structures. It uses polymers and composites in many parts. None of this is unique to EVs, but EVs change the mix and the intensity.

When people argue about "one material," they often oversimplify. The battery chemistry matters, but so does the manufacturing process. The motor design matters, but so does the sourcing of components. The most honest approach is to treat materials as a portfolio. Different designs use different portfolios. That is why supply risk differs across models and brands. It is also why the industry keeps evolving.

Supply risk is not only about scarcity. It is also about concentration and geopolitics. A material can be abundant globally and still risky if refining capacity is concentrated. A material can be available and still risky if transport routes are fragile. A material can be mined responsibly in one place and poorly in another place. So "supply risk" includes operational risk, political risk, and reputation risk. Companies manage this through supplier diversification, long-term contracts, and chemistry shifts. They also manage it through recycling and recovery plans, but those plans take time to scale.

Ethics in supply chains is difficult because it depends on verification. It is easy to claim "ethical sourcing." It is harder to prove it consistently. Verification requires traceability and audits. It also requires enforcement. Even then, systems can fail. This is not a reason to give up. It is a reason to be careful with language. A responsible statement sounds like this. "We aim for traceable sourcing and audited suppliers." An irresponsible statement sounds like this. "This product is perfectly ethical." Perfection claims are usually a warning sign.

A practical way to think about supply ethics is to focus on incentives and transparency. When a company publishes supplier standards and audit expectations, it creates a baseline. When it reports progress and problems, it builds trust. When it hides details, trust falls. Also, when buyers demand lower prices without caring how costs are reduced, supply pressure increases. Ethical outcomes require system pressure in the right direction. That pressure can come from regulation, investors, customers, and industry standards.

Recycling is often presented as a simple solution. In reality, it is a developing industrial system. Recycling has pathways and limits. The pathway begins with collection. Collection is hard because batteries are distributed across many vehicles and regions. The pathway continues with safe transport and storage. That is also hard because damaged batteries can be hazardous. Then the battery must be diagnosed and processed. Processing can mean disassembly, sorting, and material recovery. Each step has cost and safety constraints. Each step requires scale to become efficient.

Today, recycling can help in several ways. It can recover materials and reduce dependence on some primary supply. It can reduce waste. It can also create a market for end-of-life services. But recycling also has limits today. Some recovery processes may be complex. Some recovered materials may need additional refining. Some designs are easier to recycle than others. Some markets lack collection networks. So it is better to treat recycling as an improving capability, not as a magic switch.

Second-life use is another idea that sounds simple but requires discipline. A battery can sometimes be used after vehicle life for less demanding roles. This can make sense when safety and testing are strong. It can also make sense when economics work. But it is not guaranteed for every pack. Packs age differently. Their histories differ. Their safety states differ. So second-life is a case-by-case system, not a universal outcome.

Now, how do you think clearly without slogans. Start by avoiding extreme claims. Avoid "EVs are always clean" and avoid "EVs are always dirty." Both are too simple. Replace slogans with conditions. Ask where the vehicle is built. Ask how long it is used. Ask how the electricity is produced where it is charged. Ask how the battery is cared for. Ask what end-of-life plan exists. These questions do not require perfect answers. They require honest answers.

You should also separate short-term and long-term effects. In the short term, building EV supply chains can increase demand for certain materials. That can stress systems. In the long term, improved recycling and chemistry shifts can reduce some pressures. But the long term is not automatic. It depends on investment and policy and execution. Saying "it will be solved" is not enough. You need to ask "how will it be solved, and by whom."

A final clarity tool is to focus on what you can control. If you are a consumer, you can choose a vehicle that fits your needs without oversizing. Oversizing means buying more battery and weight than you use. Oversizing can increase material intensity per useful mile. If you are a fleet manager, you can plan vehicle utilization and lifespan. High utilization can improve the value per unit of material. If you are a business leader, you can prioritize supplier transparency and design for repair and reuse.

Common mistakes in this topic are predictable. One mistake is arguing from identity rather than evidence. Another mistake is treating one study or one headline as the full truth. Another mistake is ignoring the time dimension. Another mistake is ignoring the grid and focusing only on the vehicle. Another mistake is ignoring manufacturing and focusing only on operation. Another mistake is pretending uncertainty means "anything goes." Uncertainty means you use careful language and avoid overconfidence.

Quick Check

1. Why does focusing on one material often oversimplify the EV debate?

2. What is the difference between scarcity risk and concentration risk?

3. Why is traceability harder than it sounds in supply ethics?

4. Why does recycling depend on collection and logistics, not only technology?

5. Why is second-life use not guaranteed for every battery pack?

6. What questions replace slogans with clear thinking?

7. What is one practical way to reduce material intensity as a buyer or fleet?

Chapter 21 — Lifecycle Thinking and Carbon Accounting (High Level)

Lifecycle thinking means looking at the full journey of a vehicle. It starts with materials and manufacturing. It continues through years of driving and charging. It ends with recycling, reuse, or disposal. Carbon accounting is one tool used within lifecycle thinking. It aims to estimate emissions across stages. But carbon accounting has uncertainties. It can also be misused in marketing. So we will treat it as a helpful map, not as a perfect measurement.

The first key concept is the manufacturing footprint versus the operating footprint. Manufacturing footprint includes mining, refining, battery production, and vehicle assembly. Operating footprint includes electricity generation used for charging over time. A conventional vehicle typically has lower manufacturing emissions but higher operating emissions due to fuel combustion. An EV often has higher manufacturing emissions, mainly due to battery production, but can have lower operating emissions depending on the electricity source. This is a conceptual pattern. It is not a universal guarantee. It depends on location and usage.

The second key concept is the grid mix. Grid mix means the sources of electricity used in a region. Some regions use more fossil fuels. Some regions use more renewables. Many regions use a mix that changes by season and time of day. This is why location matters. Charging in a cleaner grid region can reduce operating emissions faster. Charging in a dirtier grid region can still reduce emissions compared to fuel in many cases, but the margin can be smaller. Because grids change over time, lifecycle results can change over the vehicle's life. That is why "one number for all places" is unreliable.

The third concept is vehicle lifetime and utilization. If a vehicle lasts longer, its manufacturing footprint is spread across more miles. That can improve lifecycle results. If a vehicle is used very little, manufacturing impact per mile becomes higher. This is why fleet use can sometimes show strong lifecycle performance. Fleet vehicles often have higher utilization. But fleet results depend on many factors, including charging strategy and maintenance. Again, it is a system outcome.

Battery degradation is also part of lifecycle thinking. Degradation means the battery slowly loses capacity and performance over time. Degradation is influenced by heat, charging habits, and usage patterns. Degradation affects range and usability. It can also affect resale value. But lifecycle thinking is not only about emissions. It is also about resource efficiency. If the battery lasts longer, fewer replacements are needed. That reduces material pressure. So battery care becomes a sustainability practice, not only a personal cost practice.

Now consider battery second life in this lifecycle map. Second life means a battery pack, or its modules, may be used after vehicle service in stationary roles. Stationary roles can be less demanding than driving. They can tolerate lower energy density. They can value stability and predictable output. Second life can extend the useful life of materials. It can also reduce the need for new storage products in some contexts. But second life requires testing and safety assurance. It requires business models and warranties. It requires logistics and standardized interfaces. So it fits as an emerging option, not a default assumption.

There is also the end-of-life stage. End-of-life is where recycling and recovery become important. Recovery can reduce demand for newly mined materials in the long term. But the timing matters. When EV markets are in early growth, there may be fewer end-of-life batteries available. When markets mature, end-of-life volumes increase. That can improve recycling economics. So the system improves over time, but not instantly.

What is uncertain in lifecycle thinking, and why. Many factors vary. Manufacturing practices differ by factory and supplier. Electricity sources differ by region and change over time. Vehicle use varies by driver behavior, climate, and routes. Battery chemistry and design differ. Recycling systems differ. Data collection can be limited. Also, many lifecycle models use assumptions. Assumptions are necessary, but they must be disclosed. When assumptions are hidden, results can be misleading.

A disciplined reader looks for what is measured and what is assumed. A disciplined reader also avoids comparing two lifecycle numbers without checking context. Two studies can disagree because they used different boundaries. One study may include battery recycling credits. Another may not. One may assume a certain grid intensity. Another may assume a different one. Neither is "lying" necessarily. They are modeling different realities. The correct response is not anger. The correct response is to ask what the assumptions were.

A helpful mental model is to treat lifecycle performance as a range, not a point. It is often better to say "likely lower in this region under typical conditions" than to claim "exactly X percent lower." Exact numbers require specific data and a clear time window. Without that, precision becomes fake confidence. Fake confidence harms trust.

How can you use lifecycle thinking in real decisions. Start with fit. Choose a vehicle that fits your daily needs rather than the biggest battery available. Then focus on charging behavior. If you have access to off-peak charging, you can reduce grid stress and sometimes reduce emissions depending on the region. If you can use cleaner electricity sources, your operating footprint may improve. If you care for the battery to extend life, you improve resource efficiency. None of this requires perfection. Small disciplined choices compound over years.

Businesses can also apply lifecycle thinking in procurement. They can evaluate vehicles by use case and utilization. They can design charging schedules to reduce peaks. They can set maintenance plans that extend life. They can track real energy use and uptime. They can also plan end-of-life handling. This shifts sustainability from marketing into operations. Operations is where sustainability becomes real.

Common mistakes include treating lifecycle as a weapon in debates. Another mistake is focusing only on tailpipe emissions and ignoring manufacturing. Another mistake is focusing only on manufacturing and ignoring grid improvements over time. Another mistake is assuming the grid is static. Another mistake is ignoring the effect of vehicle lifetime. Another mistake is ignoring how much driving behavior affects energy use.

Quick Check

1. What is the difference between manufacturing footprint and operating footprint?
2. Why does grid mix make lifecycle results location-dependent?
3. Why does vehicle lifetime change impact per mile?
4. How does battery care connect to sustainability outcomes?
5. Why is second life an option but not a guaranteed outcome?
6. Why can two lifecycle studies disagree without either being dishonest?
7. What is one practical way to apply lifecycle thinking as a buyer or fleet?

Part 8 — Future Trends (Practical Forecasting)

The EV future will not arrive as one straight line. It will arrive as a set of scenarios. Some places will move fast. Some will move slowly. Some will leap forward, then pause. Forecasting is most useful when it stays practical. It should focus on signals you can observe. It should also separate what is likely from what is uncertain. This part gives you a disciplined way to think about the next decade without hype.

We will not promise exact timelines. Timelines depend on technology, costs, grid readiness, policy, and consumer trust. We will instead build a scenario lens. You will learn what trends have strong momentum, what trends face constraints, and what uncertainties can shift outcomes. The goal is to help you prepare, not to entertain.

Chapter 22 — Future of Electric Vehicles (2026–2035)

Some changes are likely because they are driven by strong incentives. Other changes are uncertain because they depend on breakthroughs or political choices. A calm forecast starts by separating these categories. It also starts by admitting that the EV industry is not one industry. It is batteries, software, charging, grid operations, manufacturing, and consumer behavior all at once. When many systems must align, forecasting requires humility.

What is likely. Battery cost and performance will likely improve in incremental steps. Improvements can come from better manufacturing yield, better pack design, and chemistry optimization. Even without "miracle batteries," steady engineering progress can reduce cost and improve reliability. Charging infrastructure will likely expand in many regions because demand pushes investment. Software platforms will likely become more integrated with charging and navigation because users demand simpler experiences. Fleet adoption will likely continue where utilization makes economics clear and predictable. Quality discipline will likely become a stronger competitive weapon as markets mature.

Charging will likely become more about reliability than about raw speed claims. Many users care less about peak charging power and more about predictable sessions. They care about working stations, clear pricing, and easy payment. Operators that deliver uptime and simple access can win trust. Vehicles that charge smoothly and communicate clearly can reduce anxiety. This is a "trust trend," not only a "hardware trend."

Grid interaction will likely become more visible. As EV penetration rises, utilities will care more about peak demand. Managed charging will likely grow because it reduces infrastructure stress. Time-based pricing may expand in some markets, but details vary. Workplace and depot charging will likely become more structured because fleets can schedule and control loads. Home charging will remain important where homes can support it. In dense cities, public and shared solutions will matter more.

What is uncertain. Breakthrough chemistry that changes energy density and safety dramatically is uncertain in timing. It may happen, but it is not guaranteed within a decade. Large-scale vehicle-to-grid adoption is also uncertain. Technical capability may expand, but business models, warranties, and grid rules can slow adoption. Full autonomy is also uncertain. Driver assistance will improve, but broad autonomous operation depends on safety validation and liability frameworks. Policy support is uncertain because politics changes. Incentives can rise or fall. Regulations can tighten or loosen. Global trade conditions can also change supply chains.

There are also uncertainties in consumer trust. One major incident or repeated reliability failures can slow adoption. Strong safety performance and transparent incident response can accelerate trust. Trust is not purely rational. It is shaped by stories and personal experience. This is why companies that invest in calm communication and reliable service can influence adoption beyond pure technology.

Signals to watch can be grouped into six areas. The first is batteries. Watch manufacturing scale, defect rates, and safety practices. Watch how warranties evolve. Watch how quickly new chemistries move from pilot lines to stable mass production. The second is charging. Watch uptime, repair times, payment simplicity, and interoperability progress. The third is the grid. Watch managed charging programs, transformer upgrades, and the growth of depot charging. The fourth is policy. Watch emissions rules, safety and cybersecurity expectations, and charging infrastructure support. The fifth is software. Watch update discipline, cybersecurity incidents, and the quality of user experience. The sixth is supply chain resilience. Watch diversification and recycling scale.

Now we build three scenarios for the next decade. These scenarios are not predictions. They are planning tools. A good planning tool helps you act even when you do not know the future.

Scenario one is steady acceleration. In this scenario, costs fall gradually. Charging reliability improves. Grid upgrades keep pace. Policy remains mostly supportive or at least stable. EV adoption grows across segments. Fleets grow quickly where utilization is high. Consumer adoption grows where home charging is common and public charging improves. Competition shifts from "who can build an EV" to "who can deliver the best system." In this world, winners are those with operational excellence, strong service, and trusted charging partnerships.

Scenario two is uneven growth and fragmentation. In this scenario, some regions become EV-dominant while others lag. Charging improves in major corridors but remains inconsistent elsewhere. Supply chain shocks occur periodically. Policy changes by region create complexity. Many brands exist, but user experience varies widely. Fleets adopt where economics are clear. Consumers adopt in pockets where infrastructure works. In this world, the biggest advantage is ecosystem fit. Companies that tailor to local realities do better than those chasing a single global model.

Scenario three is trust-driven slowdown and reset. In this scenario, adoption slows due to public trust concerns, reliability issues, or infrastructure frustration. It may also slow due to economic conditions or policy reversals in some regions. Progress continues, but buyers hesitate. Companies respond by improving safety, service readiness, and transparency. Charging operators focus on uptime. Standards and enforcement become stronger. After a reset period, adoption can resume with stronger foundations. In this world, the winners are those that treat safety and reliability as core products, not as marketing.

What survives in each scenario. The stable winners are systems that reduce friction and risk. Home charging solutions where feasible will remain valuable. Depot charging for fleets will remain valuable. Reliable fast charging on key routes will remain valuable. Clear warranties and strong service networks will remain valuable. Transparent battery health reporting may become more important because it supports resale and trust. Cybersecurity discipline will remain necessary because connected vehicles are long-lived devices.

A practical takeaway is that "technology alone" is not the main bottleneck. The main bottleneck is system readiness. System readiness includes infrastructure, safety, service, and trust. Companies that build system readiness can survive volatility. Buyers who evaluate system readiness can make better decisions without being misled by the newest headline.

Common mistakes in forecasting include chasing the most exciting breakthroughs and ignoring boring constraints. Another mistake is treating the future as one global outcome. Another mistake is assuming adoption is only about price. Another mistake is ignoring grid constraints. Another mistake is ignoring human behavior and trust. The most disciplined forecasts combine engineering, economics, and public acceptance.

Quick Check

1. What is one trend that is likely without requiring breakthroughs?
2. Why is charging reliability often more important than peak speed claims?
3. What makes vehicle-to-grid adoption uncertain in practice?
4. Why is policy a major uncertainty for EV forecasting?

5. What is one signal you can watch that relates to trust?
6. What is the main difference between the three scenarios described?
7. What "system readiness" factor do you think matters most in your region?

Conclusion

A Calm Summary: What You Know Now

You now have a system map of electric vehicles. You understand the vehicle as energy storage, power conversion, control, charging, grid interaction, and lifecycle. You understand why range is a trade-off, not a promise. You understand why charging is a service system, not a plug. You understand why software adds value and also adds responsibility. You understand why regulation and standards shape adoption as much as technology. You understand why sustainability requires lifecycle thinking, not slogans.

How to Stay Safe and Smart

Stay calm and stay disciplined. Use reputable charging equipment and follow manufacturer guidance. Treat software updates as part of safety, not as entertainment. Keep your habits simple and consistent. Do not chase extreme claims or fear-based stories. When you see a headline, ask what is known and what is assumed. When you compare products, compare system fit for your use case. When you hear a promise, ask what must be true for it to hold.

Your Next Steps (learning + practice + discipline)

Pick one use case and study it deeply. If you are a buyer, study your charging access and driving pattern. If you are a fleet manager, study depot charging, uptime, and maintenance training. If you are a business leader, study supply chain resilience, service readiness, and trustworthy communication. Keep a short checklist for clarity. Update your understanding slowly and carefully. Progress in this field rewards patience more than excitement.

Glossary (A–Z)

AC Charging — Slower charging that uses alternating current, often at home or work.

Adapter — A connector piece that lets one plug type fit another port. Use only approved ones.

Aero Drag — Air resistance that increases energy use, especially at highway speeds.

Aftermarket Charger — A non-original charger. Quality varies, so choose reputable brands.

Air Cooling — Using air to remove heat from batteries or electronics. It is simpler than liquid cooling.

Anode — The battery electrode where lithium is stored during charging.

Availability — The share of time a charger or system is working and usable.

Auxiliary Battery — A small 12V battery that powers lights, locks, and computers.

Battery Capacity — How much energy a battery can store. Larger capacity often means more weight.

Battery Degradation — Gradual loss of capacity and performance over time.

Battery Management System (BMS) — The "brain" that monitors and protects the battery pack.

Battery Pack — The full battery unit in the vehicle, made from many cells.

Battery Thermal Runaway — A dangerous chain reaction where a cell overheats and can ignite.

Bidirectional Charging — Charging that can send power both into and out of the vehicle.

Brake Blending — Mixing regenerative braking and friction braking for smooth stopping.

Busbar — A metal strip that carries high current inside a battery or power unit.

Cathode — The battery electrode that receives lithium during discharge.

Cell — The smallest battery unit. Many cells form modules and packs.

Charge Curve — How charging speed changes as the battery fills. It is not constant.

Charge Port — The physical inlet where you plug in to charge.

Charge Rate (C-rate) — A measure of charging speed relative to battery size.

Charging Network — A set of charging stations operated under one brand or system.

Charging Session — One continuous period of charging from start to stop.

Charging Standard — A shared technical rule for connectors and communication.

Charger Uptime — How often chargers are working when drivers arrive.

Charger Utilization — How much a charger is used over time. High use can stress equipment.

CHAdeMO — A DC fast-charging standard used in some vehicles and regions.

Combined Charging System (CCS) — A connector system used for AC and DC charging in many markets.

Connector — The plug shape and pins that transfer power and data.

Contactors — High-power switches that connect or disconnect the battery safely.

Cooling Loop — Tubes and fluid paths that move heat away from batteries or electronics.

Current (Amps) — The flow of electricity. Higher current can mean more heat in cables.

Cycle Life — The number of charge-discharge cycles a battery can handle before major wear.

DC Fast Charging — High-power charging using direct current, often for travel.

DC-DC Converter — A device that converts high-voltage battery power to 12V power.

Demand Charge — A utility fee based on peak power use, important for fast-charging sites.

Degradation Curve — The pattern of battery aging over time, often faster early then slower.

Depth of Discharge (DoD) — How much of the battery is used before recharging.

Derating — Automatic power reduction to prevent overheating or damage.

Drivetrain — The system that moves the vehicle, including motor and gears.

Drive Unit — A packaged motor, inverter, and gearbox module in many EV designs.

Efficiency — How much energy becomes motion rather than heat loss.

Electric Range — The distance an EV can drive before needing to recharge. Real range varies.

Energy (kWh) — The total stored or used electricity. Think of it as the "fuel amount."

Energy Density — Energy per weight or volume. Higher density can reduce pack size.

Energy Management — Software strategies that control heating, cooling, and power use.

ESC (Electronic Stability Control) — A safety system that helps prevent skids, present in many cars.

EVSE — Charging equipment that supplies power safely to the vehicle.

Fast Charger — A charger designed for high power, usually DC.

Firmware — Low-level software that controls hardware behavior.

Fleet Depot Charging — Charging many vehicles at one site under managed schedules.

Frunk — Front storage space in some EVs due to different packaging.

Gear Reduction — A fixed gear ratio used to match motor speed to wheel speed.

Grid Mix — The sources of electricity in a region, such as gas, solar, or wind.

Ground Fault — Unwanted current path to ground. Systems detect it for safety.

Grounding — Connecting electrical systems safely to earth reference in installations.

Heat Pump — A system that moves heat efficiently for cabin or battery heating.

High Voltage (HV) — The main battery system voltage, usually hundreds of volts.

HV Interlock — A safety loop that disables power if connectors are opened.

HV Junction Box — A box that distributes high-voltage power to major components.

ICE (Internal Combustion Engine) — A traditional engine that burns fuel.

Incentive — A policy support tool like rebates or tax benefits. It varies by region.

Inverter — Converts battery DC power into AC power for the motor.

IP Rating — A protection rating against dust and water for components and connectors.

Isolation Monitoring — Checking that high-voltage circuits are safely insulated from the body.

Kilowatt (kW) — A rate of power. It describes how fast energy flows.

Kilowatt-hour (kWh) — A unit of energy. It describes how much energy is stored or used.

Level 1 Charging — Very slow AC charging from a basic household outlet in some regions.

Level 2 Charging — Faster AC charging using dedicated equipment.

Liquid Cooling — Using liquid to remove heat from batteries or power electronics.

Load Balancing — Managing charging power to avoid overloading a site or transformer.

Losses — Energy wasted as heat in wires, electronics, or motors.

LFP — A battery chemistry often valued for safety and long life, with lower energy density.

Lithium-ion — The common battery family used in most modern EVs.

Market Penetration — The share of sales or vehicles that are EVs in a region.

Managed Charging — Controlling charging times and rates to support the grid and reduce cost.

Motor Controller — Hardware and software that controls motor torque and speed.

NACS — A connector standard used by some networks and vehicles in certain regions.

NVH — Noise, vibration, and harshness. EVs change NVH patterns due to quiet motors.

Off-Peak Charging — Charging when grid demand is lower, often cheaper and easier on the grid.

Onboard Charger — The vehicle device that converts AC from the wall into DC for the battery.

Open Charge Point Protocol (OCPP) — A common protocol for charger-backend communication in many networks.

Over-the-Air (OTA) Update — Software updates delivered through the internet without a workshop visit.

Peak Demand — The highest power draw in a time period. It drives grid and cost planning.

Permanent Magnet Motor — A motor type that can be efficient but uses rare-earth magnets.

Plug-in Hybrid (PHEV) — A car with a battery and engine, able to charge from a plug.

Power (kW) — The rate of energy use or delivery at a moment.

Power Electronics — Devices like inverters and converters that control electrical power.

Powertrain — The system that creates motion, including battery, inverter, motor, and gearbox.

Preconditioning — Heating or cooling the battery before driving or fast charging.

Range Anxiety — Worry about running out of charge before reaching a charger.

Range Estimator — The software prediction of remaining distance. It depends on conditions.

Reciprocity — Network agreements that allow one account to use multiple charging operators.

Regenerative Braking — Using the motor to slow the car and send energy back to the battery.

Reliability — Consistent performance without failure, especially important for charging networks.

Reserve Buffer — A hidden energy margin to protect the battery and avoid deep discharge.

Road Load — The forces the vehicle must overcome, including drag and rolling resistance.

Rolling Resistance — Tire and road friction that uses energy, especially at lower speeds.

Rooftop Solar + EV — Using solar power to support home charging, depending on system size.

Route Planner — Software that plans charging stops based on battery and charger availability.

Second Life Battery — Reusing a battery for stationary storage after vehicle use, when feasible.

Self-Discharge — Slow loss of charge while parked, usually small but not zero.

Service Network — The repair and support ecosystem that affects ownership experience.

SOC (State of Charge) — The battery's current charge level, similar to a fuel gauge.

SOH (State of Health) — A measure of battery aging and remaining capacity compared to new.

Solid-State Battery — A battery concept using solid electrolyte. Timelines and scale are uncertain.

Standardization — Agreement on interfaces and tests to reduce friction and improve safety.

Supercharging — A branded term for high-speed charging, not a universal standard name.

Thermal Management — Systems that control heat in batteries, motors, and electronics.

Thermal Runaway Propagation — Heat spreading from one cell to others, increasing risk.

Time-of-Use Pricing — Electricity prices that change by time, influencing charging schedules.

TCO (Total Cost of Ownership) — Full lifetime cost, not just purchase price.

Telemetry — Operational data sent from vehicle or charger for monitoring and diagnostics.

Transformer Capacity — The local grid limit that can constrain site charging power.

Trip Efficiency — Real energy used per distance on a trip, affected by speed and weather.

Ultra-Fast Charging — Very high-power charging categories. Experience still depends on battery and curve.

Vehicle-to-Grid (V2G) — Sending power from the vehicle back to the grid under control.

Vehicle-to-Home (V2H) — Using the vehicle to power a home during outages or peak periods.

Warranty — The manufacturer promise for repairs, often with special battery terms.

Winter Range Loss — Reduced range in cold weather due to heating and battery chemistry effects.

WLTP — A test cycle used in some regions to estimate range and efficiency. It is not real-world exact.

www.ingramcontent.com/pod-product-compliance
Lightning Source LLC
Chambersburg PA
CBHW052208220526
45471CB00004B/1867